# THE BIRD TABLE BOOK

# THE BIRD TABLE BOOK

## How to Attract Wild Birds to your Garden

TONY SOPER

ILLUSTRATED BY
ROBERT GILLMOR

d&C
David & Charles

*Page 2* Great Tit on peanut feeder. *R. Bird/FLPA*
*Page 3* Blue Tit in flight.
*Page 5* Long-tailed Tits in the hedgerow.

*Other books by Tony Soper*
Birdwatch
Owls (with John Sparks)
Penguins (with John Sparks)
The Shell Book of the Shore
A Passion for Birds
Oceans of Birds
The National Trust Guide to the Coast

A DAVID & CHARLES BOOK


First published 1965
Sixth edition (revised) 1992
Reprinted 1994
All bird and wildlife artwork by Robert Gillmor.

Artwork on pages 94–5, 122, 127, 167, by Robert Gillmor, reproduced courtesy of the RSPB.



A catalogue record for this book is available from the British Library.

ISBN 0 7153 0053 9

Typeset by ICON, Exeter
and printed in Italy by Milanostampa SpA
for David & Charles
Brunel House Newton Abbot Devon

# CONTENTS

# INTRODUCTION

Birds come in most shapes and sizes. With the benefit of millions of years of research and development, they have learnt to take advantage of every food source offered by both the plant kingdom and their fellow animals. Some have specialised in living close to man, thus becoming our mess-mates and sharing the crumbs from our table. Many more enjoy the fruits of our labour indirectly, finding food and shelter in landscapes which are largely created to our specifications, while others suffer as a result of those same drastic changes to the environment. The object of this book is to explore some of the ways in which we may exert our influence to improve the quality of bird life and to enjoy a closer relationship with those birds which prove amenable to manipulation.

Enticing garden birds to the bird table for food and to nestboxes for breeding is highly entertaining for us and, from the birds' point of view, a profitable arrangement. The procedure is by no means confined to the traditional milieu of suburban gardens; wherever there are birds to be found there are ways of improving facilities in such a way that both their lives and our own are enriched.

As time goes by, more and more species are learning to visit garden feeding stations and take advantage of man's activities and generosity. In Devon, for instance, forty different species have taken food from a rural garden. Reed Buntings, Siskins and Great Spotted Woodpeckers are examples of birds which are increasingly welcomed in suburban gardens in many parts of the British Isles. Exotic birds like the Hoopoe, visiting southern England in the summer when they overshoot the channel, are as often as not seen in private gardens, enjoying the insect harvest of a well-kept lawn.

But while rarities are of interest, far and away the greatest pleasure of bird gardening is the year-in, year-out companionship of a group of individuals which have thrown in their lot with us and become residents. The Swallows and House Martins of summer are a joy to welcome, but the Robin which feeds from the hand and sings through the winter brings a deeper sense of community with his belonging.

Many crumbs have been eaten by many Robins since this book was first launched twenty-seven years ago. And interest in bird gardening continues to grow, as my postman will confirm! Each new edition reflects the changing character of the bird table scene. When the first *Bird Table Book* was written, no Siskin had ever taken a peanut, so far as we know. Now they are two-a-penny on

Not all British birds are small and brown! These are all colourful garden visitors. Jays and Green Woodpeckers are year-round residents, Waxwings arrive in hard winters and Hoopoes in hot summers. Rose-ringed Parakeets are cagebird escapees, sustained by suburban bird tables.

RG

mesh baskets, joining Reed Buntings and Lesser Spotted Woodpeckers, and also Rose-ringed Parakeets and other exotic creatures.

It is clear that more species are learning to take advantage of bird tables and feeding stations. Just as well, when so much of their natural feeding ground is being swallowed up for unsympathetic development. All the more reason to try to create a house and garden which takes into account the requirements of wildlife as well as your own.

Feeding birds (and Hedgehogs, Badgers and toads, too) is a rewarding activity. Not only are these creatures good to see about the garden, but their relationships with each other, with Man, and with their surroundings, are of absorbing interest. So spare them an honest crust, with a bit of cheese as well.

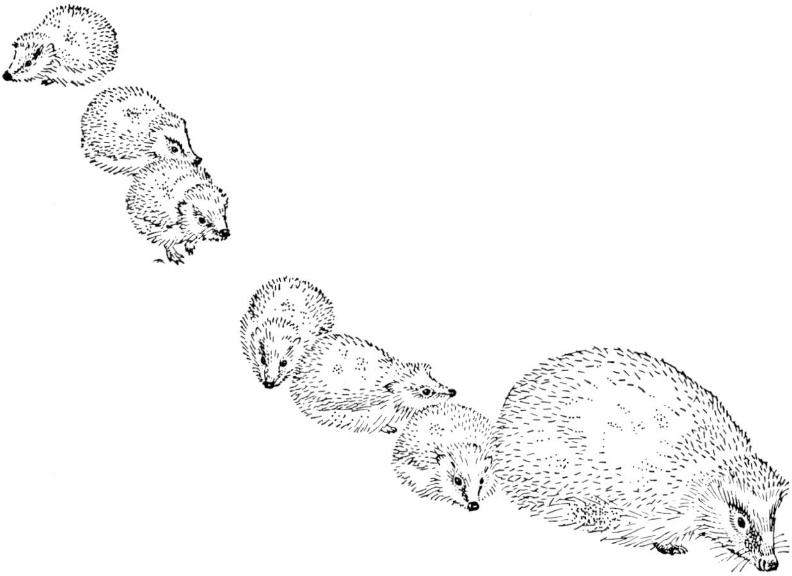

# 1

# BIRD GARDENING

## HISTORICAL BACKGROUND

On the grand scale, bird management involves the farming of large acreages in order to provide favourable conditions for resident or visiting species, thus increasing their numbers. The pioneers of this sort of activity worked in America, where suitable lakes and lakeside vegetation are still strictly controlled for the benefit of migratory wildfowl. In Britain, the Royal Society for Protection of Birds (RSPB) manages a network of bird reserves which, at the same time, serves the purposes of other wildlife. The productive and healthy farming of any habitat will support the liveliest population of birds, so what's good for worms is good for birds, too, in the long run.

It seems likely that Saint Cuthbert, living the simple life at his hermitage on Inner Farne, off the Northumberland coast, in the seventh century AD, ran the first bird reserve in the world. His was an all-embracing love, encompassing Otters and seals as well as Eiders and Gannets, to say nothing of his fellow men. He enjoyed all his birds, but had a special affection for the Eiders, affording them total protection from the hunters. His idyllic group of islands is still a wildlife sanctuary, under the protection of the National Trust.

Inevitably, the earliest records of active bird management derive from their economic value. Some centuries ago pigeons were encouraged to breed in convenient coastal caves, for example at Wemyss in Fife (although the practice originated much earlier in the Middle East), in order to take advantage of the resulting fat squabs. And while wildfowl had been hunted and herded in northern countries (during their flightless period of moult), the first account of managed breeding herds in Britain comes from Abbotsbury in Dorset. Here the monks of the Benedictine Monastery encouraged the nesting Mute Swans, taking a proportion of the fat cygnets for the table. The first written records of this activity were in 1393. The swans did well on the rich feeding in the coastal lagoons of the Fleet, inside Chesil Bank. In 1543, Henry VIII granted the right to keep the herd to the Fox Strangways family, who have protected the swans through the years to the present day. The swanherd watches over a flock which numbers some four hundred to five hundred birds at breeding time, but rises to nearer one thousand in winter.

It was Squire Charles Waterton, of Walton Hall near Wakefield in Yorkshire, who deliberately set out to create a bird park of his property, and thus he probably became the proprietor of the first nature reserve in Britain to be established for pleasure and enjoyment. In 1817, he began a ten-year plan which systematically developed his 260 acres into a sanctuary designed to improve the prospects for birds. He banned shooting altogether, threatening to strangle his gamekeeper if he shot any Barn Owls, and forbade boating on his lake during the waterfowl breeding season. He also built a high wall round the entire property to exclude casual disturbance, and dogs. Foxes were trapped and deported, and even Badgers were shown the door, in rather drastic moves to reduce the hazards to birds. Curiously, though, he encouraged Weasels, inveterate small-bird killers, but this was because he had an obsession with ridding the estate of the abundant Brown Rats, the 'Hanoverian' rats which he hated.

Waterton's enclosed reserve, and his management of it, was highly successful. In a letter to his friend George Ord, dated 1849, he wrote '...my carrion crows, herons, hawks and magpies have done very well this year and I have a fine brood of kingfishers. They may thank their stars that they have my park wall to protect them. But for it their race would be extinct in this depraved and demoralised part of Yorkshire.'

He experimented with various kinds of nestbox designed to encourage owls (and Starlings!) to breed. He ensured, too, the luxurious growth of Ivy, correctly

Walton Hall, Yorkshire seat of Squire Waterton, probably the first scientifically managed nature reserve in Britain.

deciding that it was a wholly beneficial plant which would provide food and shelter for the birds without damaging his trees. One way or another Squire Waterton was far in advance of his contemporaries in understanding the general principles of wildlife management on an ecological basis, in spite of his aberrations. He successfully sued the owners of a nearby soap works when he considered their effluent was polluting his lake, but the law, reflecting the spirit of an age when industry was of key importance, awarded him derisory damages. Yet interestingly, Waterton's relations with fellow workers in his field were inclined to be fairly hot-tempered. He cordially despised the museum workers, calling them 'closet naturalists' (this was at a time when taxonomy, the science of classification, was seen as the only fit pursuit for a professional), and positively gloried in field-work as well as his amateur status. When he shot birds for his collection (he was an enthusiastic taxidermist) he chose only males, because he preferred the bright colours! However, in an age when estates were kept up primarily for the production of game, and keepers were almost without exception wedded to the concept of vermin destruction with little understanding of the relationship between species, Waterton's ideas were regarded as highly eccentric, an epithet which he firmly rejected but invited time and time again by indulging in a series of bizarre exploits. One of his many party tricks in later life was to scratch the back of his head with the big toe of his right foot. He must bear his share of responsibility for the long-standing belief that naturalists are not entirely stable creatures.

By the end of his time Waterton had recorded 122 species of birds in his park, including Osprey, Hobby, Hooded Crow and Crossbill. Sadly, after his death the

Spotted Flycatchers take over a flower basket; ideally it should be hidden away in dense honeysuckle and there should be a convenient perch a few feet away. *Brian Hawkes/NHPA*

estate was bought by a man who 'improved' it back to conventional tidiness and ruined Waterton's life's work.

## AIMS OF THE BIRD GARDENER

Large scale bird management tends to be connected with agriculture, or sport (in the field of game conservation). But there is plenty of potential for influencing the lives of birds so that their lot is improved and our enjoyment of their company increased. And, for this process, the term 'bird gardening' is a convenient one, whether the area in question is around a factory, a school playing-field, or the garden of a house. Man's works inevitably attract their quota of wildlife, just as inevitably as they deprive some species of their living space – they not only diminish, but provide opportunities. The bird gardener seeks to enhance these possibilities. After all, the birds will come whether we like it or not, so we might as well enjoy them. And by providing new possibilities we may reverse the trend towards impoverishment in terms of species diversity.

Quite apart from their practical usefulness in terms of pest control and as agents of seed dispersal, birds are important to us as a food source. But not only are they useful, they are beautiful as well. Their colour and movement, their calls and songs add whole new dimensions to existence. So the diligent bird gardener sets out to please them and to organise a welcome. Put simply, this means providing food, water and shelter.

Food is the prime requirement of any animal. As fuel, it provides energy, promotes health, allows for defence capability and the capacity for reproduction. Water is equally important, not just for drinking but as a necessary requirement in the maintenance of plumage. Shelter provides protection from enemies and the elements, and serves as a nursery. A long-established, mature garden with a diversity of shrubs and flowering plants, with a mix of young, prime and decaying trees, some clad in ivy, an orchard with lichen-encrusted apple trees and surrounded by thick impenetrable hedges, allied to a house with plenty of corners and ledges, to say nothing of cracks and crevices, is already a most productive bird haven. The highest bird density in Britain is found in suburban gardens and estates, where the habitat resembles an endless woodland edge, one which lacks the less 'birdy' forest interior. But it has to be recognised that this habitat doesn't suit all species, so special attention needs to be given to those which have been dispossessed by house building. This is why the bird gardener also supports the intensive work done by organisations like the RSPB.

Faced with the prospect of a newly built house, surrounded by a morass of subsoil, you must obviously survey the situation carefully before rushing into a planting programme. First of all, make sure that your precious topsoil is carefully replaced. Ensure, also, that any existing boundary walls or hedges, and other features such as trees, ponds or wells, are carefully preserved, so that you are able to write them into your plan. Determine the soil type, consider the orientation of the ground, and cast a long look at the biological opportunities that abut onto your property. While your neighbours may jealously guard their territorial rights, wilder animals will use the highways which suit their purposes without any consideration for property law. The object is to reinforce the positive advantages of your own patch, to introduce complementary possibilities, and to exert what influence you can on the neighbouring land. If there is a public park nearby, for instance, it may be worth encouraging the authorities to improve their tree-planting plans by concentrating on native trees, thus discouraging those dark and dismal laurel/rhododendron plantations.

In the garden, the object should be to plant for maximum nut, berry, seed and insect production, to provide a measure of cover, and to provide water. In a newly planted garden, which will inevitably lack abundant bird food, a bird table will be most important, but the long-term plan should be to provide as varied and plentiful a mix of natural food as possible. Make the place an adventure course for enterprising bird explorers. A varied terrain of lawns, rockeries, walls, miniature hills, valleys, streams and ponds. If the visiting birds find a rich cornucopia, they are more likely to move in and settle.

The process starts with an abundance of greenery, providing food for insects which in turn support Blue Tits, which in their turn support Sparrowhawks. The very nature of gardening itself, the process of turning soil and planting new life, encourages a particular group of birds – those which are best able to adapt to our activities. Robins, thrushes, sparrows, Starlings and Dunnocks are all predisposed to live alongside us and will soon take advantage of new housing estates. On the other hand, many of our summer visitors, such as warblers, need to be encouraged, by providing a wealth of foliage and flowers which support quantities of insects.

## A Variety of 'Weeds'

Seriously consider the possibility of keeping a wilderness area in part of your garden, a wild jungle of weeds and shrubs which can be visited by hunting bands of itinerant finches. A clump of nettles will be a hothouse breeding-ground for insects and spiders, and the leaves will serve as egg laying sites for butterflies. If possible, the wilderness area should have a dark and secret roosting-place where birds may rest and recuperate, but the most important factor is a flourishing variety of seed-producing 'weeds'.

Nettles are good value – and beautiful, too. Other suitable plants are thistles, knapweeds, Teazel, ragworts, Groundsel, chickweeds, Dandelion and docks. Although these sound like a veritable catalogue of disaster, the advantage of these native plants is that they offer first-class feeding opportunities to our birds which are well equipped to exploit them. Goldfinches use their long probing bills to extract the seeds from prickly thistle and Teazel heads. It is true that wild thistles have an unfortunate tendency to run riot, but the ornamental varieties, which are most restrained territorially, still produce plenty of seeds. Cow Parsley, that vigorous and glorious hedgerow edge plant, should be a welcome member of the wilderness community and Greenfinches much enjoy its seeds. And the same is true of Fat Hen, a plant cordially disliked by 'real' gardeners but equally cordially enjoyed by finches when they go for the seeds in late summer.

Bramble should find a place somewhere. Apart from providing good roosting and nesting potential, its flowers support insects, Comma caterpillars feed on its leaves and, in due course, thrushes and Blackbirds take the berries, voiding the pips which in their turn are found by passing finches.

## TREES AND SHRUBS

Obviously, your ability to nurture trees will depend on the size of your plot, and it is true that plenty of low, dense cover is far more important in a small garden. If you can possibly find room for it, grow one tall tree – for instance a poplar – which will serve as a song post for a thrush. But even if denied a natural song perch, your thrush or Blackbird will happily accept the second-best option of a chimney pot or television aerial. And if you are lucky enough to have an old decaying tree on your patch, then cherish it for any number of reasons which will become apparent. Felling should only be regarded as a last resort, in cases of potential danger. Enjoy the well-grown trees you have; plant new ones for your grandchildren.

Where you have space to plant new trees, choose native species by preference rather than the exotics which nurserymen will be only too glad to recommend. The foliage, fruits and seeds of native trees will be more efficiently harvested by our birds, which are programmed by long experience to utilise them to best advantage. Be careful to consider their eventual size when you are planting specimens which seem puny at the time. So allow space, but allow for thinning.

Our native trees are better able to withstand our moist and wayward weather, and harmonise well with the other plants and animals in their community. Mature oaks and limes support a flourishing community of their own, but of course they do take time to grow. 'Two hundred years a-growing, two hundred a-thinking and two hundred a-dying' just about sums up an oak's life, but for all that people think of it in terms of slow growth, it can reach a goodly girth and height well within one man's lifetime. In forest conditions an oak may reach up 130ft (40m), but a lime may exceed that by yet another 30ft (9m). The lime, however, is very amenable to pruning, and as one of the few forest trees to be pollinated by insects, it hums with activity in summer. One particular aphid chooses limes on which to enjoy sap and exude its sticky honey-dew, attracting queues of bees and other insects. So naturally it is a favourite with birds, which also enjoy its autumn fruits.

If you are looking for trees which will grow fast and provide a quick return for garden birds, choose ash, elm, birch, willow and native cherry. Ash grows rapidly in any soil, and its seeds – keys – are an important food source for Bullfinches and others. Birch, too, will grow fast in most soils, but it is somewhat disease-prone, an advantage perhaps for the bird gardener who lives in the hope

Like all thrushes, the Mistle Thrush is highly appreciative of windfall fruit, especially in hard weather. *A. R. Hamblin/FLPA*

of providing woodpeckers with an easily drilled home. Its seeds are eagerly taken by Redpolls, Siskins and tits. For smaller gardens there is a dwarf version, *Betula pendula youngii*, a weeping birch.

The purple berries of Elder are enjoyed by dozens of species of birds, so it clearly merits a place in any birdman's garden. The native form is preferable to cultivars; it grows fast, almost anywhere, and it is hardy, taking plenty of punishment. Since its leaves appear early in the season it provides valuable nesting cover for our native songbirds, which breed earlier than the migrant visitors.

Ideally, trees in a garden should exhibit a mixed age structure, with young trees allowing plenty of light to reach the ground plants and prime trees providing an abundance of food and shelter. In the long run, decaying trees are the most valuable of all, allowing living space and providing sustenance to the greatest variety and number of insects and plants which live off their bounty. If you find the spectacle of a slowly dying tree a trifle uncomfortable, then clothe it with ivy to make it look more interesting!

Fruit trees in an orchard supply large quantities of bird food, both in terms of insects and, more controversially, the fruit itself. But it may be possible to leave some of the fruit on a few of the trees, so that it will decay gently into the kind of soft flesh for which thrushes are so grateful in the winter. In summer, Blue Tits will hunt over apple trees and take quantities of the Codling Moth caterpillars which cause so much damage. Both apples and pears suit birds very well, as do most other fruits. Wild Cherry is a satisfactory bird tree, but avoid the sterile double-flowered cultivated varieties. Blackbirds and Starlings, too, will be pleased to help you harvest redcurrant and flowering currants. Of course, it is easy to object to the way birds take their tithe of fruit and table vegetables, but the other side of the coin is a valid one: Starlings, for instance, eat large numbers of leatherjackets.

Spindle is a useful shrub, growing as much as 15ft (4.6m) high. Bushy and ornamental, it prefers chalky or lime-rich soil. In autumn it sports attractive colours, after producing a rich crop of pink and orange fruit. However, this fruit is poisonous to humans and should not be planted if you have young children. But Rowan (Mountain Ash) is a first-class bird tree: it needs plenty of light but is not fussy, although it prefers a light soil. Fruiting in August, its brilliant coral-red fruits are a magnet for Mistle Thrushes, Blackbirds, Song Thrushes and Starlings, which will strip the tree long before winter. A useful bonus with Rowan is that it will provide protection for your property against the evil designs of any passing witch. And if the extensive Rowan berry crops of Scandinavia fail, then in winter eastern Britain sometimes enjoys an invasion of the spectacular Waxwings, birds which depend upon the berries and hunt rather desperately through our hedgerows in search of substitutes.

Hedges are very important to birds, providing endless opportunities – for both food and shelter. So it is worth spending some time and effort on growing satisfactory ones. If possible, the ideal is to mix the plants so that they provide a range of food choices which peak at different times of the year. Hawthorn, for example, makes a good basic choice. As a free-growing tree it will grow quite tall, but it bows gracefully to life as a disciplined hedge – cut to shape, layered and trimmed it provides dense cover. It grows quickly, almost anywhere, and its spiny branches soon form an excellent boy-proof barrier. It presents lovely colour in spring, while its secret interior houses songbirds' nests. From August it provides a generous crop of scarlet haws, luscious berries which are taken by thrushes and Blackbirds, as well as winter visitors like Fieldfares and Redwings, to say nothing of Waxwings. Be careful not to confuse it with Blackthorn, whose sloes are not much liked by birds.

Holly is another first-class hedge plant, although it is reluctant to fruit when it is hard clipped, and it prefers well-drained soil. It may be a slow grower, so make a point of buying vigorous (and expensive) stock from the nurserymen. If you are free-growing it for berries, make sure you plant females, but there must be one pollinating male nearby to be sure of effective fertilisation. The cultivated forms are most reliable. 'Golden King' (a female) grows to 10ft (3m) and crops well; 'Madame Briot' to 18ft (5.5m), producing golden berries. As a hedge plant, Holly mixes well, providing a good evergreen cat-proof hedge with an impenetrable roosting and nesting fastness. Birds are not enthusiastic about the berries, except in hard weather, but the Holly hedge pays its way by virtue of its secure winter roosting potential.

Hazel is a useful addition to the hedgegrow, on the grounds of diversity, in that it provides a welcome nut harvest in August and September. It flourishes best on rich, chalky soils which are not too wet. Find a sunny space or two for some Crab-apple trees in the hedge, as they are an invaluable source of winter food for thrushes in hard weather. Fieldfares and Redwings will enjoy the flesh, leaving the pips for finches. The fruit resembles outsize yellow cherries and is reluctant to fall, even after the leaves have been shed, thus providing good food very late in

## BIRD TREES FOR THE GARDEN

*(With maximum heights)*

**Forest giants** – don't plant these unless you have a giant garden!

Ash 80ft (24.4m) *Fraxinus excelsior*. Popular look-out and song posts. Ash keys provide winter food, especially for Bullfinches.

Beech 100ft (30.5m) *Fagus sylvatica*. Beech mast is *the* major winter food for many garden birds. Massive form and total shade.

Oak 80ft (24.4m) *Quercus robur, Q. petraea*. Single most important wildlife tree – fruits and insects for feeding, abundant nest sites. Huge spreading form.

Hornbeam 80ft (24.4m) *Carpinus betulus*. Greenfinches and Hawfinches love winter seeds. Unfortunately Hawfinches prefer trees in groups, rather than singly!

**Deciduous natives**

Alder 60ft (18.3m) *Alnus glutinosa*. Wet sites. Insect rich and woodpeckers, tits, Treecreepers, warblers and Redpoll, Siskin and others feed on winter seed cones.

Aspen 70ft (21.3m) *Populus tremula*. Quick-growing, insect rich – suits warblers and tits especially. Light shade; attractive trembling leaves.

Birch 60ft (18.3m) *Betula pendula; B. pubescens*. Another 'top drawer' bird tree: insects and seeds. Redpolls and others feed on spring catkins. Woodpeckers and Willow Tits nest in decaying stumps. Quick growing; light shade.

Elder 20ft (6m) *Sambucus nigra*. Not often considered for planting but a good wildlife tree. Flowers attract many insects, birds love autumn elderberries. Flowers and berries are winemaker's delight!

Hazel 20ft (6m) *Corylus avellana*. Hazel nuts are important winter food for birds and mammals. Another insect-rich tree: good bird feeding.

Hawthorn 20ft (6m) *Crataegus monogyna, C. oxycanthoides*. Boy-proof hedgerow plant but also makes a good standard. A key bird tree for all-year insects and winter berries. Much used for thorny nest sites.

Larch 80ft (24.4m) *Larix decidua*.Use for quick-growing shelter on a bare site. Good seed producer.

Rowan and Whitebeam 50ft (15.2m) *Sorbus aucuparia, S. aria*. Attractive white flowers, light shade, heavy autumn berry crop. Quick growing; both produce berries early in life.

Wild cherry 70ft (21.3m) *Prunus avium*. Grows to very big tree. Good nest sites; popular summer fruit crop for birds.

Willows, *Salix spp*. Many species – heights vary; all first class for birds. Avoid larger species – Pussy willow, *Salix caprea*, has everything to commend it – size, shape, amenable to pruning, earliest flowers and masses of seed in autumm. Warblers, tits, Goldcrests make a bee-line for it.

**Native evergreens**

Holly 50ft (15.2m) *Ilex aquifolium*. An important tree – nest sites, roosting, insect food and mid-winter berries. Berries only on female trees; male tree needed for pollination.

Scots Pine 80ft (24.4m) *Pinus sylvaticus*. Too big for most gardens. Good for corvid nests; owls. Rich insect fauna; pine cones for Crossbills.

Juniper 15ft (4.6m) *Juniper communis*. Good native with popular berries for birds and good nesting shelter. Innumerable varieties now available, choose to suit needs, but ensure it will have berries.

**Non-native evergreens** – speedy short-term shelter.

Lawson Cypress 60ft (18.3m) *Chamaecyparis lawsoniana*. Tightly columnar. Many varieties: *'Elwoodii'* (bluish) is good for small gardens. Nesting for Dunnock, thrushes, Goldfinch. Good roosting and pinnacle song post for Chaffinch and Greenfinch.

Leyland Cypress 50ft (15.2m) *Cypressocyparis leylandii* Fastest grower bar none. Rapid shelter and nest sites. Good wind-proof hedge; cut to height. Castigated as suburban take-over plant but has good bird uses.

the winter. There are various ornamental versions but it is probably best to stick to the native Wild Crab-apple *Malus pumila*, though an ornamental variety 'Gold Hornet' has been recommended for its abundant crop of small yellow fruits, much appreciated by thrushes. Tits work hard to get at the pips, and Chaffinches take the pips after the fruits have been hacked open by thrushes. The variety 'Veitches Scarlet' produces large, scarlet fruits.

Yew is another useful hedgerow plant. It will suffer endless clipping and live a long and fruitful life as a 6ft (1.8m) bush. (Though if you let it, it will live a thousand years and grow to 90ft (27.5m).) The foliage and bark, as well as the seeds, are poisonous to domestic stock so it is not suitable for farm hedges. However, it serves well in a bird garden. The evergreen hedge provides useful nest-sites and the fleshy red berries of the female tree provide good feeding for thrushes and Starlings which eat the pulp and pass on the poisonous seed without harm. Remember, incidentally, to introduce males into the hedge, though in some instances both male and female flowers appear on the same tree. Of course, a close-clipped hedge plant will not fruit as generously as a free-growing tree, but a well-varied boundary of evergreen and deciduous fruit, together with nut-bearing plants is a decided advantage from the birds' point of view. Do your hedging and ditching, tree lopping and any necessary felling early in the year, certainly by the end of March, and then the birds will not be disturbed at nesting time.

## CLIMBING PLANTS

Trees, whether they are free-standing or part of a hedge system, may be much improved by teaming them with climbing plants. And if considerations of space, or respect for your neighbour's view, mean that you cannot have trees at all, then at least erect some trellis or fencing to provide something for a climber to conquer.

Honeysuckle is a liana which will entwine and climb, providing a colourful display of early flowers and heady fragrance, but in the process will allow for some secret nest-places. Its nectar is attractive to Privet Hawk-moths which reach it with long probosci; Blackbirds, tits and Blackcaps are not enthusiastic but will take the berries. *Lonicera fragrantissima* grows bushy and is suitable as a sunny hedgerow plant; *L. periclymenum*, the native woodbine, is the entwiner and prefers some shade.

The much-maligned and ill-treated Ivy should be treated with respect and cherished in any birdman's garden. Castigated by the ignorant as a strangler of trees, it in fact does no harm except in the very rare cases when it completely covers the crown and cuts off light to the foliage. It is also frequently said that Ivy sucks the goodness out of any tree which it climbs. This is untrue: Ivy takes no nourishment, neither does it restrict the rate of growth of its 'host'. The fact is that Ivy is a top-class birdman's plant. Thriving even in poor soil, it will carpet the ground till it finds an opportunity to climb. The dense, leathery leaves do not have the fragile beauty of any number of exotic imports, instead they are a robust

Treecreepers like to build their nests behind loose bark on a tree trunk, but any dark corner or crevice will do. *John Hawkins/FLPA*

part of the British garden and woodland scene. Climbing by virtue of the deceptively root-like hairs on its stem, Ivy flowers when it reaches light. Flowering late, in September and October, it offers rich nectar at a time when this scarce commodity is particularly appreciated by butterflies, bees and other insects. Similarly, Ivy fruits exceptionally late, in March and April, when its berries supply desperately needed late food for Wood Pigeons and thrushes, to say nothing of a number of small mammals. Aside from this virtue, which should guarantee Ivy a welcome in every garden, its convolutions and evergreen secret places provide a rich source of nest sites and roosting places. Through the year, its hairy stems and nourishing leaves support quantities of insects, which in turn provide food for hunting Wrens and tits. The 'Irish Ivy' is a good cover for north-facing walls. A rampant grower, it is dense and grows out from the wall in bush shapes, providing not only exceptionally good roosting places but free thermal insulation for the house as well.

## EVERGREENS

Apart from food and, for some species, water as well, one of the important bird functions of greenery in a garden is to provide roosting shelter at night, especially in winter. So, as well as Ivy, it is worth making sure that you have a fair proportion of evergreens in your garden, for instance in the boundary hedge. Even laurels and rhododendrons are useful in this respect. Town councils like them because they shade the ground and inhibit the growth of weeds, but they nurture few insects and offer little food value. Both Common and White Spruce make useful roost trees and offer good nest-sites as well. If you are lucky enough to have a Wellingtonia (and if you plant one remember it grows to be the biggest tree in the world – in its native California it is known modestly as the 'Big Tree'), cherish it as the preferred roost tree for Treecreepers. They excavate an egg-shaped burrow in the spongy fibrous bark, then repair to it for the night, tucking themselves into the hollow in a vertical posture, bill resting on the bark and tail down, feathers fluffed out boldly – a most astonishing sight. Wellingtonia was introduced to Britain in 1853 (the year after the Duke of Wellington died). Before this time Treecreepers made hollows in rotting trees, or utilised natural cavities or crevices behind loose bark, as of course large numbers of them still do. But it is relatively easy to discover them on Wellingtonia, sometimes as many as a dozen or so, low down on the same tree. Leyland Cypress *Cupressocyparis leylandii*, for all that it is much derided, has great value in a bird garden for the dense shelter it provides for roosting and nest sites.

## DECIDUOUS AND EVERGREEN SHRUBS AND CLIMBERS

Various forms of *Cotoneaster* are useful, as ground-huggers, bushes and climbers, because they are insect-rich as well as providing a cornucopia of berries. *C. integerrimus* fruits conveniently late, between the hawthorn and Ivy harvests. Thrushes, finches and tits enjoy its rich red berries and it grows almost anywhere. Useful cultivated forms include:

C. *simmonsii*. Effective hedge plant, produces good bird berries. Height to 12ft (3.6m).
C. *horizontalis*. Deciduous. Fan-shaped cover for wall. Height up to 10ft (3m) if against wall. Red berries last well into winter.
C. *bullatus* 'Cornubia'. Height and spread 6ft to 15ft (1.8m to 4.6m). Deciduous. Clusters of red berries on arching branches.
C. *conspicua*. Sometimes known as C. *wardii*. Grows 4ft to 6ft (1.2m to 1.8m) Grey foliage and orange berries.
C. *dammeri* and C. *prostrata*. Small, ground-hugging, ever spreading trailers. Good ground cover encouraging bugs.
C. *buxifolia*. Evergreen wall climber, to 5ft (1.5m).

C. *franchetii*. Evergreen wall climber, to 10ft (3m).
C. *lacteus*. Evergreen wall climber to 15ft (4.6m).
C. *aldenhamensis*. Slowly grows to small tree.
C. *watereri*. Slowly grows to small tree.

Like *Cotoneaster* the barberry also shows itself in a bewildering variety of forms – full species and hybrids. But all of them offer rich pickings for birds, as well as colourful foliage which provides thick ground cover inhibiting weed growth. *Berberis vulgaris*, the Common Barberry, is a deciduous plant, branching thickly and growing to 6ft (1.8m), providing weed-free ground underneath. It is easy to grow, and can tolerate sun or shade – indeed, almost anything the British climate offers. Not soil-fussy, it is easily propagated from hardwood cuttings in September. Its coral red berries have a high vitamin C content and are eaten by most birds, especially blackbirds, which make a charming picture in late autumn when they perch among the fruits. Useful cultivated forms include:

B. *darwinii*. Evergreen, reaching and spreading to 10ft (3m). Good prickly hedge, needing only light trimming to shape. Purple/blue berries. Recommended varieties are 'Gold' and 'Flame'.
B. *aggregata*. 'Buccaneer'. Deciduous. Long-lasting berries. Reaching and spreading to 9ft (2.7m).
B. *wilsoniae*. Deciduous. Formidable thorns make it a useful hedge plant. Translucent coral berries.
B. *pruinosa*. Black fruits.
B. *thunbergii*. Height 6ft (1.8m), spread 8ft (2.4m). Berries and autumn foliage brilliant red.

The Firethorn, *Pyracantha coccinea*, is a high-intensity, evergreen berry producer, much enjoyed by thrushes. It climbs well, fruiting even on a dark north wall, producing deep red berries and growing to 15ft (4.6m). *P. atalantioides* grows to much the same height, but will take pruning and provides good nest opportunities.

If you can get it, *Coriaria xanthocapa* is a useful suckering sub-shrub which grows slowly to provide a good ground cover over several square yards. Its foot-long racemes bear translucent yellow fruit which is much appreciated by songbirds.

## COLOUR PREFERENCE

The question of colour preferences by birds is fascinating. The orange fruited versions of Firethorn seem to be preferred to the red ones, for instance. And birds seem to go for the black rather than the red berries. But it may be that the colour contrast between fruits and foliage is a factor and the degree of shininess in the fruits may also be significant. A cockshy list of preferences seems to indicate that red or crimson fruits are prime favourites, followed closely by orange and yellow. Then come the less favoured – black, white, blue, brown-purple and rose-pink. Much the same is true in the case of flower colours, with the same top preferences, reds to yellows, followed by the blues and whites. House Sparrows tend to attack red and yellow flowers in spring and autumn in order to eat buds and petals. In their defence, it has to be said that they enrich the soil with droppings which contain nitrogen, phosphate and lime. Like Blue Tits, House Sparrows also enjoy nectar, and so visit clumps of Red Hot Poker, *Kniphofia uvaria*. It might be well worth including this plant in your flower borders.

## BORDER PLANTS AND ROSES

Many of the cultivated border plants provide good feeding for birds, mostly in terms of the provision of seeds. Sunflowers, for example, when left to seed, are heavy with an oil-rich bounty which is irresistible to tits, Nuthatches and finches. Finches are seed specialists with sophisticated tastes, visiting plants as they come into production. Other plants produce pollen, nectar, flowers and vegetable material which is attractive to bees, butterflies, moths and so on, which are in turn preyed on by birds. The following are good 'bird' plants for the herbaceous border: Snapdragon, Cornflower, Forget-me-not, Michaelmas Daisy, Evening Primrose, Pansy, Cosmos, China Aster, Scabious, Common Field-poppy. The American Chokeberry, *Aronia melanocarpa*, produces brilliant autumn foliage along with huge black berries which are eagerly taken by birds.

The rose family provides pretty, scented blooms which are largely unattractive to foraging birds, though if there are good numbers of hips these may be shredded by Greenfinches looking for the seeds. The Guelder Rose, *Viburnum opulus*, produces clusters of red berries which may be reluctantly taken by birds. It grows to 18ft (5.5m), but there is a cultivated version, *V. compactum* which grows to 6ft (1.8m). The Japanese Snowball, *V. tomentosum plicatum*, is sterile and useless from a bird's point of view. The Wild Rose, *Rosa rugosa*, produces good fruits and is much to be preferred to the more modern and showy varieties which are selected for colour, shape and scent at the expense of anything which might please either bee or bird. In other words, roses are an abomination to the bird gardener, to be compared with the equally unproductive Tulips, Dahlias and similar plants which offer a waste land to birds. The general rule is to prefer native species to exotics. Buy British!

## LAWNS AND LAWN-WATCHING

An open and well-kept lawn is a priceless asset to the bird-rich garden. It provides a courting arena for pigeons, a battleground for Blackbirds, not to mention a great deal of choice food. Artificial and man-made though it may be, the green sward of the lawn, clearly observable and yet a constant attraction to bug, beast and bird, is a boon to the naturalist-gardener. Curious that this green invitation to laze and relax is the result of such brutal treatment to an inoffensive plant. The constant cutting down of its efforts to reach maturity is hard on the grass, and to add insult to injury, we cart off the cuttings to rot elsewhere, depriving the lawn of the very nutrients which aid its survival. Of course, these are replenished to a certain extent by the activity of soil bacteria and by the considerable efforts of worms, but over the years mowing will impoverish the soil if you deny it the cuttings. That is why, if you want to avoid bare patches, you need to import manures to counteract nitrogen deficiency. Again, if you cut too often and too close you will be providing conditions favourable to ground-hugging plants like Dandelions, Daisies and plantains. So give the grass a chance and keep the weeds in the shade, which doesn't suit them at all. Incidentally, the compost heap provides good pickings in that it is a hothouse of insect-food production and a highly productive wormery. In winter its warmth may keep a small area free of snow and provide a welcome hunting ground.

Lawn-watching offers the easiest way of observing some of the different techniques used by birds for feeding. One of the constant wonders of the natural world is its diversity, the extraordinary range of plants and animals in any given habitat, and the way that they all manage to make a living by occupying slightly different niches. Superficially, the 'nature red in tooth and claw' approach may seem justified, but it would be more accurate to see communities of different creatures living in tolerable harmony. Birds come in all shapes and sizes. Some hunt by day, some by night, some are vegetarian, some are meat eaters; some eat anything they can get hold of, including other birds. They may walk after their food, hop for it, fly for it, dive or swim for it. And each is specially equipped for the chosen job. One way or another anything which grows or moves gets eaten. Fruit, nuts, seeds, leaves, bark, living or decaying matter – all is grist to the mill. Very roughly we can divide birds into four categories according to the shape of their beaks: the hard-billed birds like sparrows, or finches which have nutcracker bills; the soft-billed birds, like Robins, which deal with insects; the dual-purpose bills which take on all-comers; and the hook-billed predators like Sparrowhawks.

On the lawn, the most obvious visitors are the birds searching for worms and soft grubs. The old saying about the early bird getting the worm is an exact observation of fact. Worms are creatures of moisture and mildness, early morning dew suits them, sunrise and sun warmth causes them to return underground. So thrushes and Blackbirds comb the lawn at first light, and this is when you may see the bigger Blackbird steal worms from the Song Thrush, thus getting his breakfast the easy way.

Birds have a good sense of hearing, but they hunt almost entirely by the sense of sight, and to some extent touch – at least that is true of those most active by day. When the thrush catches a worm, it does it because it has seen it first. The frequent false observation that birds 'listen' for worms is based on a characteristic human weakness: people make the classic mistake of regarding birds, or any other animal for that matter, as if they, too, were people. The worm-hunting thrush hops a few paces, then stands very still and cocks its head to one side. A pause, and then the stab. So we deduce that the bird had its head cocked to listen for the sound of the worm. But the observer failed to take note of the fact that the bird's *eye* happens to be in the position where the human *ear* is found. When a man cocks his head in that attitude, he is listening intently. When a thrush does it, it is watching intently.

There is another procedure which sometimes produces good returns for birds. If a Mole is busy at its underground activities, disturbing surface lines as it tunnels, thrushes and Blackbirds will keep station on the Mole, enjoying the worms which are displaced. This sort of activity, where one species benefits from the activities of another is known as commensalism.

The bird which is a universal favourite when it swoops onto a lawn is the Green Woodpecker, with its striking green plumage and red head. Its curious flight – a few flaps followed by a glide, with wings clasped tight to the body – and yaffling call, bring it switchbacking into the garden. So remarkable is its appearance that many people find it difficult to believe that it is a British bird at all, working on the dismal assumption that home-grown species are bound to be dull and dowdy. But British it is, and a delight to see, working over the lawn and exploring for ants and ants' nests.

The other woodpeckers, equally striking in their red-and-white livery, are less attracted to ground level, but the Green Woodpecker, with its long, mobile tongue tipped with sticky mucus, searches out larvae from their hidey-holes in crevices. It may spear out larger bugs, but ants are its speciality.

Some of the lawn visitors are looking for grass and weed seeds, and of these perhaps the most attractive is the Goldfinch. The sight of a charm of Goldfinches attacking the golden Dandelion ought to be enough to convert any gardener into a Dandelion fan. They approach them with zest, leapfrogging onto stems, landing about halfway up towards the head so that they weigh it down to the ground. Then they get to work. All finches are seed-eaters, with powerful jaw muscles and bills modified for husking. They have two grooves inside the bill which locate the nut or seed, then the tongue rotates it as the mandibles crush. The husk peels off, leaving the kernel to be swallowed. Different finches go for different seeds. A Hawfinch for example is tough enough to cope with cherry and plum stones, which take some cracking. Goldfinches, at the weaker end of the finch scale, use their relatively long, narrow bills rather as a pair of tweezers, probing deep into the seedhead.

Swallows will occasionally settle on a lawn to pick up flies if they are abundant, but most of the time they are concerned with airborne flies. Perhaps the most spectacular lawn visitor, but one you're only likely to see if you live on the south coast, is the Hoopoe. With its pinkish-brown plumage, barred black on the wings and back, it swoops onto the grass with a lazy flight. On landing it shows a remarkable crest in the shape of a fan, pink with black tips. Then it struts about, probing into the soil with its long, decurved bill. Typically, it prefers parkland, orchard and open-wooded country, but it is also found in the vicinity of houses, where it feeds on lawns and paths for insect larvae. A very few stay with us to nest, in holes in trees or buildings, and some years there is a considerable influx of them. One of the Hoopoe's most endearing traits is its tameness and tolerance of Man: even the French are fond of it and refrain from shooting. Its name derives from its voice – a low, but penetrating, sexy hoop-oop-oop. The scientific name *Upupa epops* is both onomatopoeic and charming at the same time.

2

# BIRDS AND BIRD TABLES

The most satisfying way of increasing the bird population in your garden is by growing the right kind of plants and creating as near a wild environment for them as possible. But there is a great deal of pleasure and enjoyment to be derived from providing food in the most direct manner, by setting a dining-table and serving suitable dishes. And almost any food we offer, from kitchen scraps to caviar, will be eaten by a wide variety of birds ranging from tomtits to Goshawks. By providing food you can entice the birds to show themselves more freely in places where you can watch them. And, as the availability of food controls to some extent the bird population of your garden, you will also be increasing their numbers. But providing food is not a pleasure to undertake lightly. Put out some scraps in the garden and you will very soon attract new residents. They will become dependent on your generosity, and if it fails they will be competing for an inadequate supply of natural foods. Especially in cold weather, birds may lose a lot of weight overnight, and they have to make it up again during the brief hours of daylight. Death comes in a matter of hours even to a healthy small bird, if it is without food. In hard weather the real killer is hunger, not cold.

## EARLY ENTHUSIASTS

While it is entirely possible to make out a case for feeding birds in order to improve their chances of surviving natural or unnatural disasters, the reason most of us do it is because we enjoy involving ourselves in other creatures' lives and establishing closer relationships with them. Francis of Assisi was probably the first man to be credited with feeding birds from a relatively pure sense of goodwill. After a wild youth, he repented to take a vow of poverty and to devote his life to a form of pilgrimage, helping the poor. The poor, in this context, included the brute creation, which doubtless enjoyed the Franciscan's bounty without adopting his principles of poverty, chastity and obedience.

To the best of my knowledge, the first man to set up a bird table unselfishly dedicated to the sustenance of birds was John Freeman Dovaston, an early pioneer of field ornithology. In a letter to the artist Thomas Bewick in 1825, he mentioned his 'ornithotrophe', a feeding device which he had erected outside a window, to which he had enticed twenty-three species to take food on a snowy day. Over a quarter of a century later, the Rev. F. O. Morris, author of the highly successful and enjoyable, though somewhat patchy, *History of British Birds*, 1857, wrote letters to *The Times* encouraging people to put food out for the birds. But his request fell on fairly stony ground, since the thrifty Victorians didn't believe in admitting to waste of any kind, especially in the kitchen.

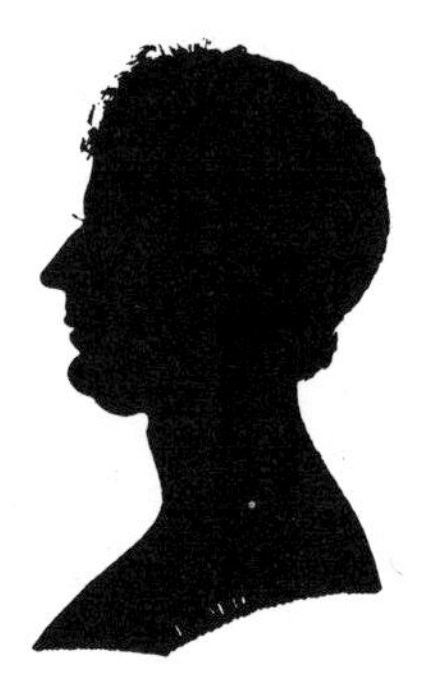

John Freeman Dovaston was probably the first man to set up a bird table. *Woodcut by Thomas Bewick.*

In Germany, at about the same time, a wealthy landowner – the Baron von Berlepsch – was pioneering techniques of large-scale bird management. His primary interest was in the control of forest pests, but the aesthetics of bird encouragement played a significant part in his thinking. Despite the fact that his main efforts went into the development of nestboxes for woodland species (as we shall see in chapter four) and that he underestimated the importance of food supply in relation to bird populations, he nevertheless experimented with the provision of artificial winter food supplies. His precise and demanding instructions for the erection and provision of 'food-houses' and 'food-bells' make fascinating reading. And he did much to encourage the spread of practical and unsentimental attitudes to the currently woolly world of bird preservation. His principal belief was that protection required an intimate knowledge of the birds' biology, and he held that Man's excesses had to be balanced by providing natural or near-natural conditions – thus von Berlepsch made it his business to lay down most precise requirements for his experiments. While he may have been somewhat over-demanding, his thinking was decidedly in advance of his times ... 'a thorough and rational protection of birds is only possible where the representatives of agriculture and forestry join forces with those who are interested in birds from aesthetic and ethical motives, and work together for a common good. Unfortunately much energy is wasted in angry quarrels.' That could as easily have been written today!

It was the long hard winter of 1890 which softened the British heart and induced large-scale bird feeding, especially in cities. By 1910, according to *Punch*, bird feeding had become a national pastime and commercial interests had begun to offer special furniture. Possibly the first person to suffer financially from the practice was brought to court during the latter stages of the First World War, in the winter of 1916-17. Sophia Stuart was charged with wasting food, when a police sergeant found a quantity of bread cut into small pieces and scattered over the ground at front and back of her house in Woking. He solemnly collected half a pound as evidence and charged the unfortunate woman. Mrs Stuart, an elderly woman who had lost her only son, claimed 'the birds are my children, I have nothing else to love', and she stoutly informed the constable that she had fed the birds for years and proposed to continue. She claimed that she only used unclean crusts, and that it was not wasted if given to one's fellow creatures whether they went on two legs or four. In spite of her efforts, Mrs Stuart was found guilty and fined £2. I hope someone offered to pay her fine.

It would make an interesting exercise to work up a defence brief for Mrs Stuart, since she could have claimed that in feeding birds with scrap food she was working for the war effort. At any time, birds fulfil a vital role in the healthy functioning of our planet's system: as insect controllers, as agents of seed dispersal, as living barometers of ecological balance and, last but not least, as primary food source. She should have claimed that she was ensuring the survival of useful allies!

The Baron von Berlepsch designed 'food houses' as an encouragement to the insect-eating songbirds in his Black Forest estates.

## IMPORTANCE OF WINTER FEEDING

Much has been made of the importance of winter bird feeding as a means of saving birds from extinction, but most of the evidence won't stand serious consideration. Bird numbers are primarily controlled by the availability of their natural foods; artificial feeding can have only a marginal influence, but it is an entirely worthwhile practice. Quite apart from the fact that it does have *some* practical effect, one reason for bird feeding is to use up kitchen scraps in a constructive manner. However, principally, the exercise gives us a lot of pleasure by attracting birds to a place where we see them to advantage. In really hard weather, extra feeding almost certainly saves a lot of individuals from an early grave. James Fisher, the ornithologist, estimated that a million birds survived one exceptionally severe winter by courtesy of bird tables. Certainly feeding sustains a higher bird population in winter, at a time when the garden is at its least colourful, and that is a prime objective. Remember, though, that this sort of feeding is at best a substitute for natural food.

Sometimes people claim that feeding birds is wrong because it interferes with the natural course of events. But the fact is that we interfere in the lives of our fellow creatures and vegetation in almost everything we do, and much of this activity is entirely proper. In any case the provision of a measure of food and water, together with a few nestboxes, represent a modest return for the loss of natural habitat we have inflicted on our wild neighbours. Taken to its logical conclusion, the 'antis' should go round knocking down Swallow and House Martin nests built under Man's own roof in order to persuade these erring creatures to find themselves a more natural cliff or cave.

One thing is certain, if you do decide to feed the birds, the best time to do it is in winter. Having started, you are honour-bound to continue until the dark days are over and your artificially maintained population is able to fend for itself in the increasingly plentiful days of spring.

In cold weather birds face several problems: the ground may be so hard that they cannot get at the invertebrate creatures of the soil; worms migrate downwards in dry or cold conditions; days are short, so hunting time is limited. Many Blue Tits, though, take advantage of street lights to work overtime and they forage almost without stopping in mid-winter. Provided birds' plumage is in good condition they are perfectly able to withstand low temperatures, but inevitably their energy requirement is increased as temperatures drop and they use fuel to keep them warm. Many birds lose ten per cent or more of their body weight overnight in cold weather, and the short daylight hours must produce food to replace the lost fat. While a large bird like a gull may manage comfortably for a couple of days, provided it can fill its belly with a decent fish from the fish quay or some high-energy waste from a rubbish tip, a Wren has a continuous appetite. Small species have a relatively larger-scale food requirement. Though physically small, they have a proportionately larger body surface and lose heat fast. Birds must maintain the highest body temperature of any animal – between 104°F and 112°F (40°C and 44.4°C), as against Man's 98.6°F (37.0°C). And chemical reactions occur more rapidly at high temperatures. One way or another song-birds must work hard and fast at feeding.

## BIRD TABLES AND THEIR SITING

The traditional way of feeding is with a bird table and, although it has limitations, it is on the whole a very satisfactory method. The table can either be supported on a post or it can hang from the bough of a tree, or a bracket. Your own situation will probably decide the method you use. There is little to choose between the two systems so long as you keep the cat problem firmly in mind and do not fix the hanging model to a potential cat-way.

Birds not only have varying food requirements, both in nature and at the bird table, but they also have varying methods of hunting. Some skulk about on the ground, some snoop along branches and foliage, and some run about on tree trunks and stone walls. So we must have variety of presentation as well as variety of food.

Ground feeders, such as Blackbirds, thrushes, Dunnocks and Moorhens prefer to feed at ground level, so they are best fed from a suitable tray which is taken in at night to cheat the rats. But put the tray a good 6ft (1.8m) from the sort of cover which might hide a stalking cat. The greatest variety of species, however, come to visit a bird tray which is fixed 5ft (1.5m) or so off the ground, in a position where most of them will merely regard it as an unusually shaped tree branch. Tits, finches and Robins will be the regular visitors, making a dozen or so everyday customers. There is great potential, too, for surprises when using this kind of tray, as we shall see later, and a fair chance of colourful visitors arriving – like woodpeckers, Nuthatches and exotic creatures from Scandinavia and the Mediterranean. Even if your only contact with the outside world is by way of a window above ground-floor level, you have a fighting chance of seducing birds to take advantage of your offerings, depending of course on the kind of greenery there is in the neighbourhood.

The tray should offer a food surface of 1½sq ft. to 2sq ft. Naturally, there should be no easy access for cats. Therefore, ideally, the post should be made from a piece of water pipe which will probably prove too slippery for them to climb; or perhaps the post can be sheathed in plastic tube which has the same effect. A rustic pole, of the sort which is all too often on sale in garden shops and the like, merely invites cats and other predators to shin up and take pot luck! The worst monstrosity on sale is the combination bird table and nestbox where any unfortunate owners of the nestbox are faced with an endless procession of callers, threatening their peace and causing territorial ructions.

If possible, the table should be protected from hot sun and driving winds; a roof is not essential, but has the advantage that it may keep off the worst of the rain. Some ornamental bird tables have the luxury of a thatched roof, which at least offers sparrows some useful nest material. If you have one of these, be careful it doesn't make access easier for squirrels and cats which may jump from a nearby vantage point and find the thatch offers a good landing grip. In truth, it is almost impossible to defeat squirrels; they can solve the most fiendish problems of access to food. The table should be sited carefully to afford good all-round vision to its visitors, while being 6ft (1.8m) or so from convenient cover. Birds will make use of convenient staging posts on their way to and from the bird table, so ensure these are available. If there are no suitable branches, provide some substitutes in the form of posts or horizontal perches.

The tray should be cleaned often, so it is important that the coaming which frames it has a few convenient gaps to allow odds and ends of crumbs to be swept away. Making a tray is easy enough, but there is little doubt that the most

Both Great and Blue Tits are enthusiastic bird table customers. *David Hosking/FLPA*

practical bird table, offering good value for money, is the one sold by the RSPB (for address see page 182). Designed first and foremost from the point of view of the birds, it *works* well, and this is, after all, exactly what the human customer is looking for.

If you make your own, check carefully that there are no sharp edges or protruding nails which might cut or damage the birds. Enclosing the feeding area

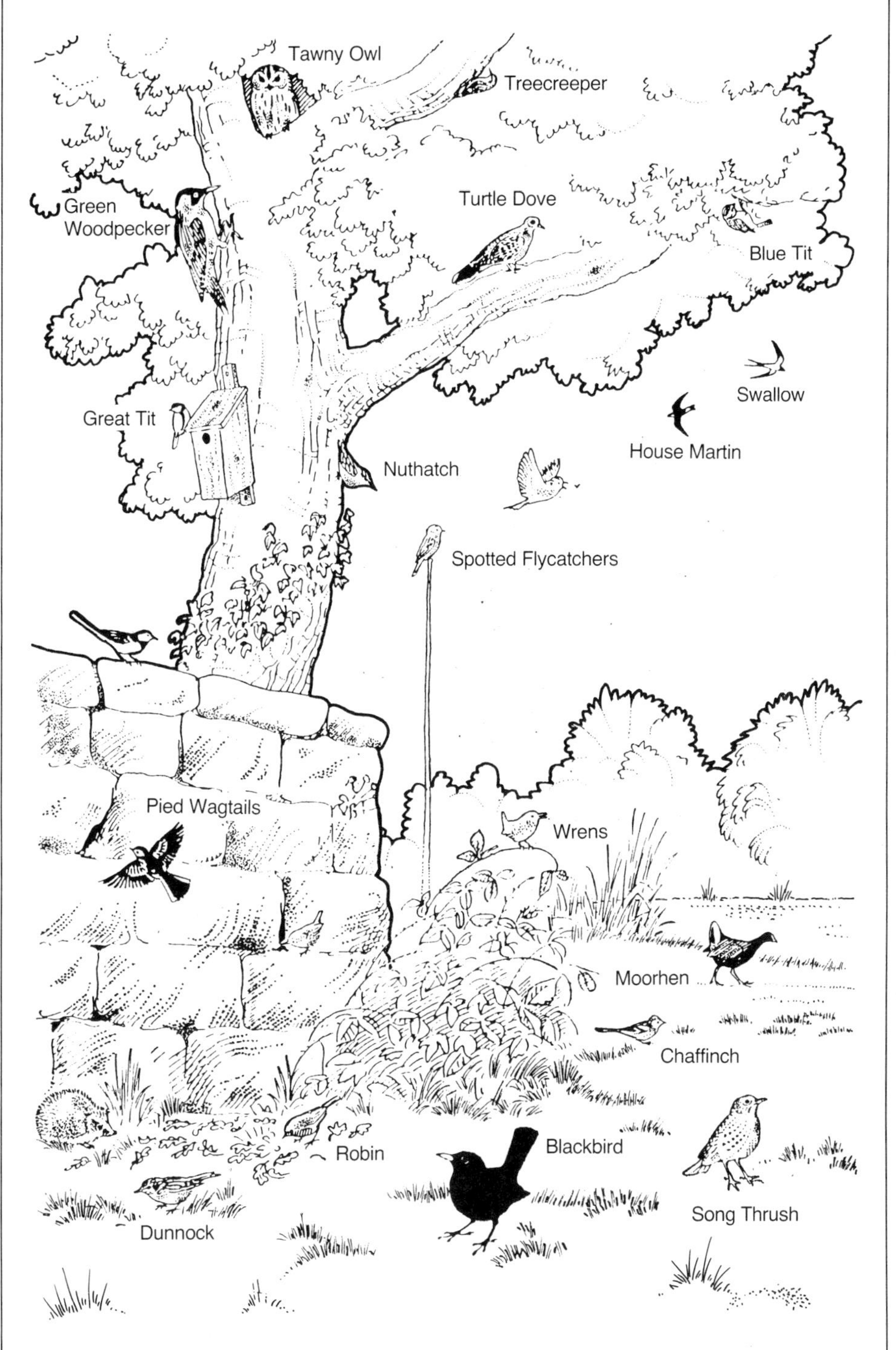


A well-planned bird garden offers a variety of habitats and attracts a variety of birds. While some are happy to show themselves on an open lawn, others prefer to skulk in the brambles or stay well above ground in a tree. Some like to hawk from a post and some are almost entirely aerial.

with chicken wire will certainly keep Starlings out, something which many people regard as desirable (not me!), but it has the undesirable effect of depriving access to thrushes, doves and woodpeckers. Probably the best solution is to provide food in a variety of ways, each allowing different birds a chance to get a share of the offerings. The RSPB sells scrap-cages, peanut baskets, seed globes and so on, and in no time at all you will start to devise your own fiendish contraptions for defeating sparrows and the dreaded Starlings. One way of beating the Starlings is to feed at first light, before they have had time to fly in from their overnight roosting place – which may be many miles away. Since birds lose weight overnight, this makes sense anyway, giving your residents a good start to the day.

House Sparrows, like Starlings, are not easy to outwit. They are omnivorous and are a successful species by virtue of the very fact that they have specialised in living as mess-mates with Man. They are tough customers at the bird table, scattering food about in a tiresome manner and ousting all non-sparrow competition. Tending to operate in pugnacious gangs, they terrorise other small birds and take what they want by virtue of their aggressiveness and sheer numbers. Thrusting and showing their muscle, House Sparrows sweep the Dunnocks, Chaffinches and tits out of the way. What can be done about this problem? Not much, but we may take advantage of one of their characteristics to give other species a short respite. They are cocky birds, and impudent – yet they are at the same time cautious by nature, suspicious of anything new. So food placed in a different position may be left untouched for other birds to enjoy undisturbed for a while. Putting food in several different places may help too. A novel food basket or bag will be safe from their attention for a few days, though they will always win through in the end.

Starlings and sparrows tend to be shyer than other birds, so if you provide a feeding patch for them well away from your house and bird table, they may patronise it in preference to the 'home' feeding stations. This is where a window-sill comes into its own, providing tits, Robins and finches with a reward for their tameness.

Not many years ago sparrows were foiled by hanging mesh baskets of the sort designed for the acrobatic tits, and much used by Greenfinches. The sparrows showed interest in them and fluttered alongside ineffectively, but failed to get a grip on the mesh. Thus, they hung about to pick up any pieces of nut which fell to the ground. As the years went by they learnt to grasp the side of the basket and pick out the nuts in Greenfinch fashion. Now this habit is widespread, another example of the learning ability that was so strikingly demonstrated by the Blue Tits when they found out how to open milk bottles. But House Sparrows are versatile performers, taking seeds from seedheads in the manner of a Goldfinch, working over trees like a woodpecker and hawking for flying insects like a flycatcher – they are pretty well invincible.

But do try to develop a more friendly attitude towards starlings. One of their problems, from a bird gardener's point of view, is that there are so many of them, and it is odd to realise that only a hundred years or so ago they were quite rare in some parts of the country. One of my correspondents, Mrs Dorothy Coomber of

Bridlington in Yorkshire, had a hen Starling which was part of the wild household for fourteen years. Recognisable because of a deformed foot, she soon became tame and was fed regularly. For many years she had brought her fledglings to be fed and, when Mrs Coomber moved to another house a short distance away, the Starling followed suit! The bird's favourite food was cheese, and it had to be *Cheshire* cheese; no other was acceptable.

## BIRD-BELL

It was the naturalist H. Mortimer Batten who made the tit-bell famous on *Children's Hour*, in the early days of radio, and for many years his wooden design was made available to garden bird enthusiasts. Sadly it is no longer produced, but the Dartmouth Pottery produces a glazed version which does the job well (for address see National Trust page 186). In essence it is a very simple device, making it possible to use up kitchen scraps in a way that allows only the most agile birds to get them, with a certain amount of difficulty. Thus the 'greedy' Starlings and sparrows are held at arm's length. The pottery bell is turned upside down and placed in a mug or bowl and thus held firmly. It is primed with scraps and seeds, has a perch twig set in it, and is then filled with hot fat, though not the sort which stays liquid at room temperature (for recipe see page 44). When the mixture has set, you hang the bell up for tits, Nuthatches and woodpeckers to explore. One of the advantages of the bird-bell is that it lasts a long time between replenishments, so it keeps your birds happy if you have to go away. Even the ground-feeders get some benefit, when small pieces get dislodged and fall down. The glazed bell is easy to clean, but the RSPB have now produced an attractive version in terracotta

### A SIMPLE SUET STICK

Bore holes in a birch log, fill them with suet and attract woodpeckers.

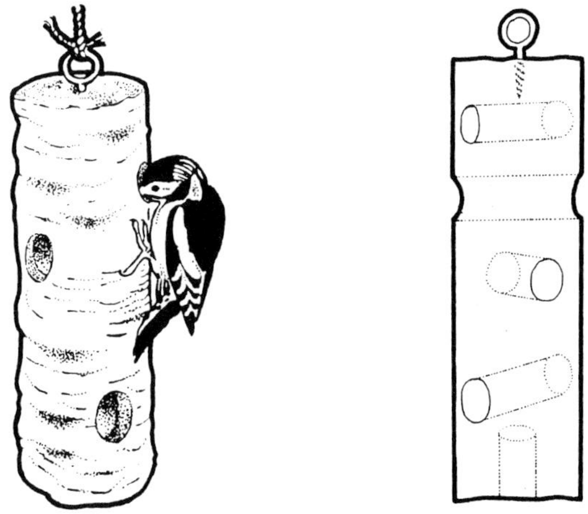

which is perhaps more 'natural'. If you do not wish to buy a bird-bell you can make a perfectly good substitute from a half-coconut, the original and cheapest version!

Another useful hanging device is the suet stick (see below). For this, bore 1in (2.5cm) holes through a short length of birch log, stuff the holes with raw beef or kidney suet and hang it up. Woodpeckers are very fond of this gadget; but avoid fancy perches, or the Starlings will take over.

The guiding principle for successful bird table operation is to offer food in an enterprising variety of ways. If the feeding station ends up looking like a Christmas tree so much the better. Quite apart from the main dish offered on the table itself, there should be hanging baskets, seed hoppers, tit-bells and anything else you may think of. The object is to give as wide a range of foodstuffs in as wide a range of dishes as possible. Avoid collapsible or bouncy spiral wire feeders though, as they may trap a bird's leg while rebounding from the shock of its arrival.

Peanuts are a favourite food, but don't provide them whole during the breeding season when they might be taken back to nestlings by inexperienced parents. *Jonathan Player/Ardea*

*Right:* Bird bells offer a useful method of providing fat-rich pickings for acrobats. A half-coconut does the job just as well as this up-market version!

*Below:* Birds' bills give a clue to their diet. Finches have nutcracker bills adapted for cracking and crushing; birds with slimmer, more pointed, bills tend to search for caterpillars and insects.

## VARIETIES OF FOOD

Different birds have different food requirements and different search patterns for satisfying them. This is, after all, the main reason that they manage to coexist so successfully. Therefore, the shrewd bird gardener studies his potential bird list, and supplies and serves food accordingly. Some species are primarily vegetarian, some are seed-eaters, some are carnivorous and some like a bit of everything. If you take a look at a bird's foraging tools you will straight away have a fair idea of what it needs. Finches have nutcracker bills, adapted to crack and crush, and they feed mostly on grain and seeds. They are hard-billed. Robins and Wrens have slender bills designed for the delicate process of probing for grubs, caterpillars

### BIRDS WHICH VISIT FEEDING STATIONS

| | |
|---|---|
| Blackbird | Moorhen |
| Blackcap | Partridge, Grey |
| Brambling | Partridge, Red-legged |
| Bullfinch | Pheasant |
| Bunting, Cirl | Pigeon, Wood |
| Bunting, Corn | Pigeon, Rock |
| Bunting, Reed | Rail, Water |
| Bunting, Snow | Redpoll, Lesser |
| Chaffinch | Redwing |
| Chiffchaff | Robin |
| Crossbill | Rook |
| Dipper | Shelduck |
| Dove, Collared | Siskin |
| Dove, Rock | Sparrow, House |
| Dunnock | Sparrow, Tree |
| Fieldfare | Sparrowhawk |
| Firecrest | Starling |
| Goldcrest | Swan, Mute |
| Goldfinch | Thrush, Mistle |
| Goose, Canada | Thrush, Song |
| Goose, Pink-footed | Tit, Bearded |
| Goshawk | Tit, Blue |
| Greenfinch | Tit, Coal |
| Gull, Black-headed | Tit, Great |
| Gull, Herring | Tit, Marsh |
| Hawfinch | Tit, Long-tailed |
| Heron, Grey | Tit, Willow |
| Jackdaw | Treecreeper |
| Jay | Turnstone |
| Kestrel | Wagtail, Pied |
| Kingfisher | Wheatear |
| Linnet | Woodpecker, Great Spotted |
| Magpie | Woodpecker, Green |
| Mallard | Woodpecker, Lesser Spotted |
| Meadow pipit | Wren |
| Merlin | Yellowhammer |

and other insects. They are soft-billed. Hawks have hooked bills for tearing fish. Gulls have general-purpose bills.

Birds eat an astonishing variety of items when they are available, though by and large they do have a decided order of preference. Thrushes prefer the rich meat of worms, but will take snails or fruit as second best. Blackbirds search for worms in the first light of dawn, only later resorting to the bird table. Natural food is clearly the best for them, and if possible this is what should be provided on the bird table and the ground feeding tray. Rowan berries, elderberries, crab apples, hazel and almond nuts, boiled conkers, sweet chestnuts, acorns and beechmast are all highly suitable, though it will be more convenient for your guests if you crush, or chop and grate the harder nuts.

Thrushes will gladly take your fruit straight from the tree, but may well be diverted if you are able to offer rejected fruit collected from your local fruiterer. Squash it first; it will be much appreciated in cold weather. If you can go to the expense of buying commercially prepared food, then Haith's 'Songster Food' is a great success with Blackbirds, Robins and Dunnocks. And peanuts please almost every kind of bird. Kitchen scraps are not only taken by the obvious 'general purpose' birds like Starlings but also by the specialist insect-eaters like Blackcaps, Chiffchaffs, Treecreepers, and woodpeckers. And seed-eating birds like Linnets, Corn Buntings, Lesser Redpolls, to say nothing of Tawny Owls and herons, are also attracted. All these, and many others, have patronised well-stocked bird tables. (For addresses of suppliers of bird food see page 184.)

The list of foods which birds will sample is almost endless, so try experimenting and remember that any left-over delicacies such as Stilton rind, Christmas pudding or even haggis will be gratefully received. The only things to avoid are salted, highly spiced or dehydrated foods. Make the offerings of either a size large enough to discourage birds from carrying them away to drop for the rats to find, or small enough to be eaten on the spot.

Bread is a controversial subject. Many people argue heatedly (and wrongly) that it should not be offered on bird tables. There are certainly dangers to avoid. Dry bread may swell inside the bird's crop and choke it, so be careful to make it moist (sparrows, and others, are careful to dunk dry bread before eating it). One solution is to soak the crumbs in bacon grease, which will be appreciated. And here is the good news about white bread. According to Derek Goodwin of the Natural History Museum, London, the incidence of beri-beri in London's street pigeons has been all but eliminated since they have been enjoying the modern vitamin-enriched sliced bread. Make of that what you will. Eric Simms, the well-known broadcaster and writer on birds, compiled a list of the birds which he had seen taking bread in the outer suburbs of London, and they totalled twenty-three – including Mallard; Herring, Common, Black-headed and Lesser Black-backed Gulls; Feral Pigeon; Wood Pigeon; Carrion Crow; Jay; Great, Blue and Coal Tits; Song Thrush; Blackbird; Robin; Dunnock; Starling; Skylark; Greenfinch; Chaffinch; Siskin; House and Tree Sparrows. Doubtless more will join the list. At a popular bird-feeding place by a car park in a woodland area of Humberside, Long-tailed Tits joined the queue for white bread crumbs. I once enticed a wild Whooper Swan to come to a well-known brand of sliced white over a period of weeks. So there may be a few surprises, and a lot of delights, in store for us all.

Finches present something of a problem in that they prefer seeds, and seeds are expensive and extraordinarily difficult to serve without scattering. If they are offered in a mixed variety then the birds will throw them all over the place as they search diligently for those they like best. They also need to be kept dry, of course. Seed hoppers are notoriously ineffective, but the 'Droll Yankee' range of tube-

*Above:* The 'Droll Yankee' comprehensive range of bird feeders. *Jacobi, Jayne & Co.*

*Above left:* Tube feeders are ideal for offering seeds to finches, since they are least susceptible to wind-scatter. *Tim Soper*

*Below left:* Wild Bird Trill is much appreciated on the bird table, but this is an extravagant way of serving it. Better to use a tube dispenser. *Tony Soper*

feeders works well. They are designed to dispense seed mixtures of the 'Wildbird Trill' type from a generous reservoir. One model offers Greenfinches tiers of perches so that a whole flock of birds can feed simultaneously. For address, see page 185, Jacobi, Jayne & Co.

Many of the seeds from a seed hopper find their way to the ground below. Not all are eaten by ground-feeders. Some may lie there and, in time, germinate and sprout into an exotic garden of unexpected plants. David McClintock, the distinguished botanist, carried out some fascinating experiments in which, instead of offering commercial bird food to the birds, he planted it. He was puzzled by the strange names which were given by the seedsmen, and found after much research, that seeds called 'Blue maw' and 'Dari', which are not found in botanical books, revealed themselves to be Opium Poppy *Papaver somniferum* and the annual tropical cereal *Sorghum bicolor* when they were encouraged to grow into plants.

## BIRD SEEDS REVEALED

Here is the table which crowned his efforts, first published in the periodical *New Scientist*.

| | |
|---|---|
| Aniseed | *Pimpinella anisum* |
| Blue maw | *Papaver somniferum* |
| Buckwheat | *Fagopyrum esculentum* |
| Mazagan canary | *Phalaris canariensis* |
| Chicory | *Cichorium intybus* |
| Dari | *Sorghum bicolor* |
| Gold of pleasure | *Camelina sativa* |
| Hemp | *Cannabis sativa* |
| White kardi | *Carthamnus tinctorius* |
| White lettuce | *Lactuca sativa* |
| Best Dutch linseed | *Linum usitassimum* |
| Chinese millet | *Setaria italica* |
| Japanese millet | *Echinochloa utilis* |
| Plate yellow millet | *Panicum miliaceum* |
| White millet | *Panicum miliaceum* |
| Niger | *Guizotia abyssinica* |
| Panicum millet | *Setaria italica* |
| Black rape | *Brassica campestris* |
| German rübsen | *Brassica campestris* |
| Chinese safflower | *Carthamnus tinctorius* |
| Striped sunflower seed | *Helianthus annuus* |
| French teazle | *Dipsacus sativus* |

Peanuts are almost the perfect produce for the dedicated feeder. Convenient to handle, store and serve, they are energy-packed with a high calorie content. But buy 'safe nuts', not mouldy ones which have turned yellow, as they may be highly poisonous, producing aflatoxin, a toxin which kills the liver cells and has caused the death of many garden birds. Unshelled peanuts will be eaten by a variety of species in considerable quantity. Offered freely, they will be used up at the rate of several pounds a week, therefore it is best to present them in bags or cages which force the birds to work at the job of freeing them. Strung, in their shells, they provide innocent amusement for us as we watch the acrobatic tits breaking and entering. But be careful that you don't string them on multi-thread cotton which might tangle up their feet. If it suits them, tits may perch on the bird table to haul up a string of nuts 'bill over claw', a version of the natural behaviour whereby they pull leafy twigs closer to inspect them for caterpillars.

Tits and Greenfinches are the prime customers for peanuts, but other species take advantage of the titbits which fall to the ground. Dunnocks, and on occasion Bramblings, will forage below while as many as three species of tit and Greenfinches are working above. Robins are fond of peanuts, too, and although they find great difficulty in fluttering alongside and grabbing a morsel, they can manage it. Jays, Chaffinches and, indeed, Bramblings, have all learnt to enjoy the bounty of the bag. One of the most remarkable instances of a bird expanding its range through a liking for peanuts is that of the Siskin, a small acrobatic finch more closely associated with Forestry Commission conifer plantations, where they enjoy spruce and pine seeds in spring and summer. Originally confined to the Caledonian pine forest of Scotland, they slowly extended south from the mid-nineteenth century, colonising parkland and conifer forests and reaching North Wales, Norfolk and the New Forest a hundred years later. Then, some twenty years ago, they started to come into gardens and feed on peanuts in south-east England. The habit has now spread through most of the country and the numbers wintering have increased, till nowadays we see them feeding on tideline seed debris on the Exe Estuary in Devon, for example.

Siskins display a curious preference for peanuts offered in the small red plastic netting containers used by greengrocers for packing carrots, coming readily to gardens with these bags rather than with the conventional RSPB scrap-cages. And furthermore, it has been confidently asserted that they prefer these red mesh bags to any other device. These birds are not only from the increasing British breeding population, but also from Scandinavia and the Baltic. Their numbers fluctuate widely from year to year, but are generally at their highest in March and early April just before they migrate back to their breeding grounds – either in northern Britain or across the North Sea. Those winters when few Siskins are seen probably coincide with good seed crops nearer to their breeding grounds.

**Use safe nuts:** Peanuts are an ideal food for garden birds. They are convenient, nutritious and contain a high level of vegetable fibre which aids digestion. A single Blue Tit may eat a pound of peanuts through the winter, if you give it the chance. Of the 100,000 tons of peanuts imported into Britain every year, 10,000 are destined for the bird table. It is absolutely essential to make sure that you offer 'safe nuts', which provide nourishment without endangering the birds' health. The Birdfood Standards Association acts as a watchdog for the wild birdfood trade, and to be sure you are buying safe nuts look for its 'Seal of Approval'.

Tits are the family which give the greatest pleasure to house-bound bird-watchers who simply enjoy the company of their birds. Their acrobatics are always a joy to watch, and they display a touching enthusiasm for any feeding device which offers nuts or fat. They very much enjoy coconut, which should be put out fresh, in the shell. Just saw the nut in half and suspend it so that the rain doesn't spoil the flesh.

Scrap baskets, such as the standard RSPB wire cage, are useful because they can be filled indoors at your convenience and the food doesn't get scattered about quite so freely as with the bird table. But, like the peanut bags, only certain birds will be able to feed from them. Thus, it is important to spread your largesse by way of a variety of feeding stations and devices.

Suet is another high-energy food, and it melts down well for the tit-bell use. The short variety is more appropriate than the stringy and it serves as an acceptable substitute for the fat grubs and insects which woodpeckers enjoy. Stuffed into mesh bags, or scrap baskets, or stuck into crevices and crannies on tree trunks, it will act as a magnet for colourful birds like Great Spotted Woodpeckers and welcome visitors like Long-tailed Tits. The brilliantly coloured woodpeckers are relatively recent bird table addicts, having first taken to the practice some time in the late 1950s. Now they are common visitors to feeding stations. Incidentally, like Treecreepers they are fond of uncooked pastry. If you daub suet into the crevices of an old, gnarled Scots Pine, it will attract other visitors as well as the woodpeckers and Treecreepers – for instance, Goldcrests and even Firecrests, together with wintering Chiffchaffs and Blackcaps.

Lastly, after that catalogue of feasting, it may be worth putting out a small

### SUITABLE BIRD TABLE FOOD

**Animal fats**, good for warblers, tits, Robins, woodpeckers, Nuthatches
Suet (beef best, or mutton)
Marrow bones, cracked (but *not* cooked bones, which might be eaten by dogs or Foxes, when the splinters may stick in their stomachs)
Bacon rinds (short pieces)
Chicken carcass (try hanging it from a tree)
Tinned pet food
Mealworms
Maggots
Ants' eggs
Cheese
Hard-boiled egg

**Fruit**, for thrushes, etc (fresh, dried or decaying)
Berries of all sorts

**Nuts**, of all sorts
Peanuts (not salted)
Almonds
Hazel
Brazils (for Nuthatches, jam them in a tree crevice)

quantity of grit – fine sand, gravel or even small bits of coal – to aid the avian digestion process. It makes sense to provide a small amount on, or under, the bird table.

Don't be discouraged if, after all your efforts, birds don't flock to your feast. It takes them time to adjust to a new feeding opportunity, as you will find if you move a familiar device, or paint something a new colour. Or perhaps there is a welcome abundance of natural food nearby, which will of course be more attractive, nutritious and generally health-promoting than anything you may have provided.

It is important to keep a bird table clean because there is an ever-present danger of bacterial infection when bird droppings accumulate. *Salmonella* kills birds. Use a five per cent bleach solution and move the table once or twice in a winter season to discourage a build-up of the potentially dangerous droppings. Do not allow a pile of uneaten, unwanted food to accumulate. And a warning – don't overfeed your birds. It doesn't make sense to provide great mounds of food, attracting quantities of birds and increasing the risk of salmonellosis and tuberculosis, both of which are common bacterial infections in wild birds. The object should be to provide a welcome *supplement*, but not so much as to induce dependence on bird table food at the expense of active foraging for a diversity of wild seeds, grasses and insects in a natural manner.

**Seeds**
Mixed (ie 'Wildbird Trill', from shops)
Hemp (a 'best buy' but must be kept dry – eagerly taken by Greenfinches, Bullfinches and Buntings, Nuthatches, woodpeckers)
Canary (Chaffinches)
Millet
Maize
Corn
Melon
Sunflower (heaven for Greenfinches)

**General**
Rice
Potato (boiled or baked in jacket)
Stale cake crumbs
Coconut (in shell, not desiccated)
Uncooked pastry (treecreepers like it)
Biscuit crumbs
Breadcrumbs
Oats (coarse, but raw, not offered as porridge which is glutinous and sticks to plumage and bills)

## Recipes from the *Cordon Bird* School of Cookery

### 1 Cakes and Puddings for the Bird Table

Basic Pud

Take seeds, peanuts, cheese, oatmeal, dry cake and scraps. Put them in a container, pour hot fat over the mixture until it is covered, and leave to set. Turn out onto a table, unless you have prepared it in a tit-bell or coconut holder. Rough quantities: 1lb (400g) of mixture to ½lb (200g) of melted fat.

Edwin Cohen's Pudding

*8oz (200g) beef suet*
*12oz (300g) coarse oatmeal*
*2–3oz (50–75g) flour*
*5oz (125g) water*

Mix flour and oatmeal with liquid fat and water to stiff paste. Bake in shallow pie dish to form flat cake at 350°F (175°C) for approximately one hour.

Miss Turner's Maize Cake

Mix 3oz maize meal in a bowl with equal quantities of chopped nuts, hemp, canary and millet seed. Stir with boiling water till coagulated, and add two beaten eggs. Tie tightly in a cloth and bake at 350°F (175°C) for fifty minutes to one hour.

Tim's Bird Cake

*2lb (1kg) self-raising flour*
*8oz (200g) margarine*
*a little sugar*

Mix with water and bake like a rock bun.

Anti-Sparrow Pudding

Boil together one cup of sugar and one cup of water for five minutes. Mix with one cup of melted fat (suet, bacon or ordinary shortening), and leave it to cool. Then mix with breadcrumbs, flour, bird seed, a little boiled rice and scraps, until the mixture is very stiff. Pack into any kind of tin can or glass jar. Lay the can on its side in a tree, on the window sill, or any place where birds can perch and pick out the food. The can must be placed securely so that the birds cannot dislodge it, nor rain get inside. May not fool sparrows for long, though, so don't take it too seriously.

### 2 Fillings for Bird Bells, Suet Sticks and Pine Cones

Basic Tit-Bell Recipe

Fill the upturned bell with seeds, peanuts, cheese, oatmeal, sultanas, cake crumbs and other scraps. Pour in hot fat to the brim. Insert a short piece of twig into the mix to act as a learner's perch, if necessary. Leave to harden. Turn the bell over and hang in a suitable place where small birds like Blue Tits are already accustomed to come for food. The bell works best in cold weather.

Another tit-bell mix, equally suitable for open offer on the bird table, comprises seeds, peanuts, oatmeal, cake crumbs and cheese. Put them in a container (bird-bell, baking dish), pour hot fat to cover. Leave to set. In the case of an open tray, simply turn it out onto the bird table. Bacon or sausage fat is ideal, using 8oz (200g) of melted fat to 1lb (500g) of mix, very roughly.

PINE CONE SURPRISE

Leave a large fir cone near a fire or radiator for several days so that it opens its scales. Gather the seeds to include in the mix. Take beef suet with any meat or fat trimmings. Melt it, stir in cake crumbs, hemp, millet seed, raisins, the pine-cone seeds, and anything else you think the birds might fancy. Pour the hot mix onto an opened cone, or dip the cone in, then allow to cool. Or stuff the holes of a feeding stick with the cooled mixture. Fix the cone or stick amongst the branches of a tree or tree substitute.

MAX KNIGHT MIX

Mix stale cake and fat with a few dried currants and sultanas.

Imprisoned in a 5–8in (125–200mm) wire-mesh bag, it keeps birds busy and prevents too much scatter.

THE BARON VON BERLEPSCH FOOD-TREE RECIPE

This mix was formulated by the good Baron as a means of extending the plenteous times of summer to woodland birds in winter, and provides high-energy, intensive feeding. It was to be poured, hot, onto either the branches of a live young conifer or an imitation tree made of separate branches. Since living trees promptly lost their leaves – the needles – when hot fluids were poured on them, he recommended this practice only in the sort of woodlands which could suffer the sight of an ugly and diseased tree. In more sensitive areas, he suggested using a felled tree, imported for the purpose. While still enjoying the Baron's recipes a century later, I think we may take it for granted that the birds will not fuss too much over the method of presentation, and this recipe serves well as a bird-bell mix. The Baron tells us that it is by no means necessary to keep closely to the ingredients; it is only to serve as a guide, though the chief part of the mix should always consist of hemp:

*5oz (150g) breadcrumbs*
*3oz (75g) hemp seed, whole or crushed*
*3oz (75g) millet seeds*
*1½oz (40g) sunflower seeds*
*1½oz (40g) oats*
*2oz (50g) ants' eggs*
*1½oz (40g) dried elderberries*
*Suet (beef suet), the less stringy the better*

Mix well, fill the bell and bind with hot, melted suet. Cool. Add more suet, melt and cool again. The second time of cooling produces a harder consistency. This mix offers a meal of high calorific value to birds which have difficulty in finding their preferred insect diet in winter.

Robins like mealworms a lot. *Eric & David Hosking/FLPA*

## 3 MEALWORM CULTURE (Robins will thank you for your efforts)

Take a smooth-sided container such as a large circular biscuit tin or one of those out-dated and highly unsuitable glass bowls traditionally used for unfortunate goldfish. An open top provides plenty of air, but have a wire mesh lid which will foil escape attempts.

Put a 4in to 6in (100–150mm) layer of dry wheat bran or barley meal in the bottom. Now a layer of hessian sacking. Add a vegetable layer of carrot, turnip, banana and apple skins, dry bread, raw potato, cabbage, as available – but ensure that the medium does not become too wet as it will then ferment, smell appalling and probably kill the mealworms. A good productive mixture will not smell.

Then take more hessian sacking and add more vegetable/bran layers to produce a multi-tier sandwich of mealworm delight. Introduce two hundred to three hundred mealworms (mealworms from an aviculturist's pet shop, *not* a fishing shop's maggots) and keep in a warm room.

After a few weeks the mealworms, fat and happy, will turn into creamy pupae,

Red Squirrels like peanuts, and are perhaps more welcome than the ubiquitous Grey.
*E. S. Turner/FLPA*

then into little black beetles, which present your breeding stock. They lay eggs which hatch into mealworms, and so on. Crop the mealworms in accordance with the scientific principles of MSY – maximum sustainable yield – in other words, make sure you keep a viable breeding stock. (If you want to start an empire, prepare other tins and prime them with a few bits of dry bread from an existing colony. These will carry beetle eggs.)

Some weeks later the mealworm city will be reduced to a dry powder, and you then need to renew the vegetable layers and bran. You may also find it useful to feed your worms with a slice of damp bread every week. Fold a sheet of paper into a concertina and place this on top of the mix; this will serve as a resting place for mealworms and simplify collection.

Serve the mealworms in a fairly deep, round dish, so that they cannot escape – they are surprisingly mobile. Gentles are sometimes recommended but, though I have not tasted them myself and the birds certainly like them, they are less wholesome to handle than mealworms. (Gentles are fly maggots, whereas mealworms are the larval form of the beetle *Tenebrio molitor*.)

### 4 EARTHWORM CULTURE

Earthworms, *Lumbricus terrestris,* are bisexual, each individual exhibiting both male and female characteristics. But it still needs two to start a family. Use a suitable box in a shady position. Fill with a mixture of sand, well-ground manure (which may include a generous helping of household peelings and greens), rich loamy soil and peat moss in equal parts. Water and mix well. Turn and sprinkle with yet more water every few days. Introduce your breeding stock after two weeks, when the mix has cooled to between 66°F and 75°F (19°C and 24°C). In three months you will have a powerhouse of worm production, to be culled for your ground-feeders' tray.

## FEEDING TIMES

It is probably best to feed at regular times, morning or late afternoon for instance, or better still, first thing in the morning only. If you have to go away be sure to leave freshly filled tit-bells and seed hoppers which keep the birds occupied until your return. Keep going right through the winter and remember that it is especially important to provide seeds for your finches in the spring, at a time when they are not very abundant naturally.

There has been much controversy about the advisability of putting out bird table food during the breeding season. The best informed opinion seems to be that while adult birds are perfectly capable of deciding what is good for them, it is possibly best to provide peanuts in such a way that they cannot be carried whole to the nest. In the breeding season you can also stuff peanut cages and scrap baskets with straw and feathers, dog or cat combings, short bits of cotton and bits of cotton wool. These will be welcomed and taken for nest material.

The Baron von Berlepsch recommended ladling hot fat on to a convenient conifer!

# 3

# BATHING AND DRINKING

## IMPORTANCE OF WATER SUPPLY

A healthy bird gets most of its water requirements from its food, but it still needs access to a supply of clean fresh water, partly because it will drink a little, but mainly because it bathes a lot. Birds don't sweat. If they get overheated they open their mouths wide and gape to lose heat, thus losing some moisture, but they also lose moisture by excretion and this must be replaced. Some species are adapted to a minimal water intake. Desert birds like the Budgerigar can go for long periods without drinking even though they are living on a dry diet of seeds. However, don't deprive your caged Budgies of water on this basis! Tree-living species may sip from foliage after rain, and for this reason there may be some sense in providing a drinking bowl well off the ground. Many town birds happily visit roof gutters to drink and to bathe.

Most birds drink by dipping their bills into the water, then raising their heads to allow the liquid to run down their throats. But pigeons keep their bills in the water, sucking it up and into the system in bulk, a method which is quick and leaves the bird at risk for the shortest time. Swallows, martins and Swifts will drink in flight in a shower of rain, or from the surface of the pond. And, like other birds, they enjoy bathing in rain or even flying through the spray of a lawn sprinkler. However, they will be cautious of bathing in a downpour, since the object of bathing is to wet the plumage but not to soak it.

The function of bathing is to maintain the bird's plumage in tip-top condition, mainly because of its importance in flight and thermal insulation. Birds need to bathe even in the depths of winter, since ill-kept plumage will not serve the purpose of keeping warmth in and cold out. Birds, like mammals, are warm-blooded creatures, maintaining a high and constant metabolic rate. Their body temperature is kept up by internally generated heat, as opposed to the system endured by reptiles and fishes whose body temperatures fluctuate in sympathy with that of the ambient environment.

The bathing process, in birds, is highly ritualised. First the plumage is made wet, but not too wet. (Caught in a heavy downpour of rain, birds will hunch and stretch up, so that the water runs off quickly.) Next, the excess drops are shaken off before the oiling stage begins. Twisting its tail to one side, the bird reaches its bill back to collect fatty oil from the preen gland on its rump. The oil is then

Most birds sip, then raise their heads to swallow, but pigeons, like this Wood Pigeon, keep their heads down and suck up a good draught. *E. A. Janes/FLPA*

Starlings bathing, a highly ritualised affair. *A. R. Hamblin/FLPA*

carefully smeared all over its feathers, the difficult job of oiling the head being carried out using the feet. After oiling, which is characteristically an urgent process, comes the more leisurely task of preening, the final stage of the ritual, where the bird nibbles and strokes its feathers, one by one, into shape. It then stretches and settles for a period of contemplation.

Birds do not only bathe in water. On occasions they will enjoy sun, rain, snow, smoke and dust baths, and they will even bathe in ants, a performance often given by Starlings or Blackbirds on your very own garden terrace. But all these forms of bathing relate firmly to the process of feather maintenance; they are simply strange manifestations of the pressing need to maintain plumage at peak efficiency.

Unless your garden is on sandy ground, you may like to provide a dust bath for sparrows and Wrens. The dusting-place should be well sheltered, with some cover nearby, and can consist of a couple of square feet of well-sifted sand, earth and ash to a depth of a few inches. Sprinkle the dust bath with bug powder or spray (eg Cooper's Household Insect Powder or Poultry Aerosol) every now and again, for the common good.

Feathers are an engineering marvel, precisely designed for their several tasks. They are purpose-built extensions of the skin, horny growths similar in origin to our own fingernails. Light but strong, amongst other things they provide lift surfaces which, powered by all that muscle, give the bird flight capability. Their surface area is large, compared with the weight involved, and their ingenious design allows for continuous maintenance and ruffle-smoothing. And when a part has come to the end of its useful life, after much wear and tear, it may be replaced without withdrawing the aircraft from service.

The power of flight gives birds the key to world travel. A tern may spend the summer nesting in the Arctic, then strike south to 'winter' in the Antarctic, incredible though that may seem to us. From its point of view it is simply making the best of both worlds. Not all birds use their feathers to propel them across the world. Flight has other values. Instant escape from enemies, airborne invasion of an area rich in caterpillars, or fast approach and capture of prey: all these things are possible with feathers. And different birds have different designs to fit them best for different purposes. A Swift has narrow, sweptback wings, designed for speed and aerial fly-chasing; its take-off and landing performance is poor. A Pheasant has broad, short wings, giving a powerful near-vertical take-off for instant escape, although it pays for the facility by having a low endurance, needing to land again within a short distance – far enough away, though, to keep out of trouble.

Even for flightless birds, which might at first seem to make nonsense of all those years of research and development, the wings are important pieces of equipment. A penquin's flipper may seem an un-feather-like, hard rigid structure, but it is in fact a modified wing superbly built for flying – *underwater*; the bird is a master submariner.

Wings are not only used for flying. They may be used as legs, as when a Swallow struggles to take a few shaky steps on the ground. On occasion they may be used as advertisement hoardings, their colour-reflecting surfaces being held up and displayed in order to intimidate a rival or impress a partner. The wings are then playing their part in the process of avian communication.

Feathers, too, have functions beyond providing lift and flight. Soft down feathers insulate the body and keep it warm; waterproof outer contour feathers repel rain and keep it dry. In some species feathers may help to guide flying insects into the gaping maw and in others feathers may protect the face from the

## FEATHER STRUCTURE

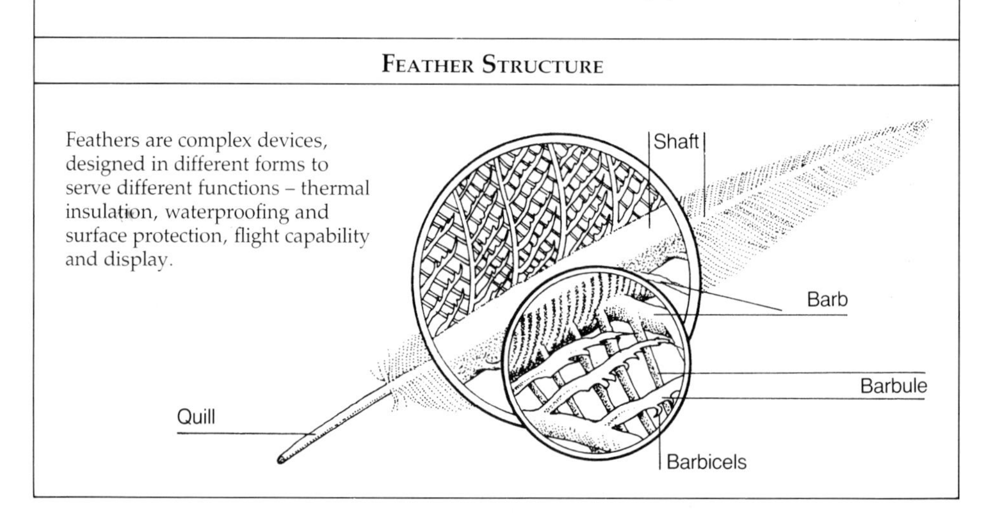


Feathers are complex devices, designed in different forms to serve different functions – thermal insulation, waterproofing and surface protection, flight capability and display.

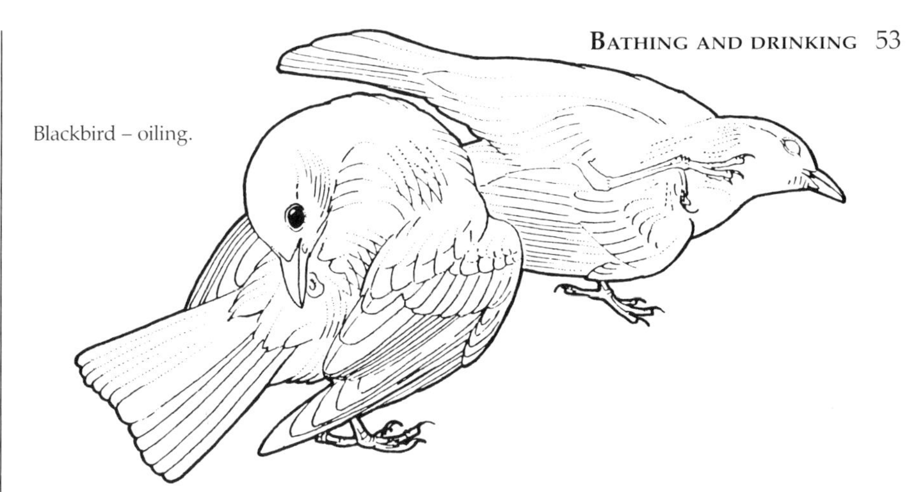
Blackbird – oiling.

stings of bees and wasps. So feather maintenance is a vital part of a bird's daily life. Much time is devoted to bathing, oiling and preening, keeping the tools of the trade in trim and keeping the bird dry, warm and ready for instant take-off. If they are badly damaged, they are replaced as part of normal growth. In the ordinary course of wear and tear they will be replaced as part of a continuous moult: a staggered process because at any given time the bird must not be at the disadvantage of having too many feathers out of action. However, ducks and geese do in fact follow a somewhat different plan, moulting all their flight feathers in one fell swoop, lying doggo and flightless for a few weeks after the breeding season while they grow a new suit for the migration flight. At this period they present a sorry appearance, but even this 'eclipse' plumage serves a purpose, camouflaging the birds at the time when they are most vulnerable.

White, or partially white, birds often appear in gardens. A survey which analysed over three thousand occurrences of this albinism showed that the six species most commonly affected were Blackbird, House Sparrow, Starling, Swallow, Rook and Jackdaw. Except in the case of the Swallow, the phenomenon affects birds which tend to lead somewhat sedentary and sociable lives. The causes are not easily defined: they may have something to do with a dietary

A partially albino blackbird.

deficiency, perhaps associated with a high intake of 'artificial' food offered at bird tables. Certainly albinism appears to be most often noticed in urban and suburban habitats, where bird table food provides a significant proportion of a Blackbird's diet.

Whatever the reason, it is a fact that the Blackbird is more prone to albinism, partial or total, than any other species. In the true albino, pigment is completely absent, even the beak, legs and eyes being colourless, but most often the condition is partial, with the plumage revealing a patch of white, or perhaps just one white feather. The extent of the whiteness may vary from season to season, and albino or part-albino young may be produced by normal parents as easily as normal young may be produced by albino parents. An individual may show more white as the years go by. Any feather on the bird may be affected, but the head is particularly prone. One can't help wondering to what extent the white Blackbird is at a disadvantage, because at the least it becomes conspicuous and therefore extra vulnerable to predation. But the suburban habitat, while it may be partly responsible for the problem, is at least a relatively protected environment. So perhaps it is a case of six of one and half-a-dozen of the other.

Whiteness is not the only genetic abnormality suffered by Blackbirds. Other 'isms' produce varying intensities of reddishness and yellowishness (erythrism and xanthochroism). In our own garden as I write this we have a gloriously honey-coloured Blackbird, an example of leucism, where the normal pigment is diluted and paled. Another common plumage abnormality is melanism, where the bird has too much of the dark pigment eumelanin. These melanistic forms have an exaggerated blackness. Pheasants provide the most commonly seen examples of this.

Whatever the colour, maintenance of a Blackbird's plumage is of great importance. At the bird bath it shows its feathers off to great advantage: the wings outspread, the handsome tail held down at a right angle and fanned out to show its full spread. Even in the depths of winter birds must bathe their plumage frequently as part of the process of keeping their insulating and flying suit in full working order. Bedraggled feathers waste body heat and make for inefficient flying, and in winter lost energy is not easily replaced.

Young Starling and House Sparrow. Juvenile plumages can be confusing – young Starlings may be mistaken for exotic thrushes; the adult's bill is yellow at courting time. *Mike Read/Swift*

## BIRD BATHS AND THEIR MAINTENANCE

Ornamental, free-standing bird baths on pedestals are poor value from the birds' point of view, though they may well improve the look of a garden – a belief which is well understood by the owners of garden centres! The sides are often too steep and the water too deep. Three inches (75mm) should be the maximum depth, and the access should be by way of a gradual non-slip slope. Birds like to wade in cautiously and do not belong to the ostentatious-plunge brigade. If you are committed to a bird bath which has a slippery, or glazed, surface, introduce some sand or gravel to make life easier for the intended users.

A long plastic plant tray makes an acceptable bird bath, the sort of thing which tends to be some 3ft 6in (1100mm) long by 8in (200mm) wide and up to 2in (50mm) deep. Or an inverted dustbin lid will serve perfect well. Either will be eagerly used for drinking and for bathing by Grey Wagtails, Spotted Flycatchers, Bullfinches, Chaffinches – as well as the usual Blue Tits, Blackbirds, sparrows and so on. But the best material is undoubtedly stone, for instance granite, or concrete.

If the bird bath contains no oxygenating vegetation it must be cleaned frequently, or there will be a build-up of algae which will stink. The water should be changed frequently (daily in hot weather), and must be kept topped up. It is best set up in shade, in the open but within reach of cover and safety.

In cold weather, the ice-free water you put out will be gratefully used. If the bath is of a breakable material, put a tennis ball in it, so that if the water freezes the ball takes the strain. Do not use glycerine or any anti-freeze, which will damage a bird's plumage. It may make sense to cut a piece of thick polythene sheet to line the bath, so that when it freezes you can flip out the ice easily, in order to refill.

In icy weather keep the bird bath ice-free by liberal use of boiling water. It may be possible to arrange a small thermostatically-controlled immersion heater, but the most elegant solution to the problem is to use a solar panel. Much experimentation has produced a workable system which is available from C. J. Wildbird Foods (for address see page 185).

## PONDS

On the whole, ornamental bird baths are not for the birds. The facilities offered by a natural or semi-natural pond are better by far. Indeed, a pond is almost an essential for any self-respecting bird garden. Properly stocked with oxygenating plants and supporting a healthy population of aquatic insects, snails and crustaceans it will provide clean water for drinking and bathing as well a useful food supply. It is also an endless source of delight for the naturalist.

A dustbin lid makes a good bird bath, if you haven't room for a pond.

There is a good reason for increasing the number of garden ponds, since they to some extent replace the gradual decline in the British pond scene. Ponds were once abundant, serving useful purposes for man as well as supporting a large number of aquatic plants and animals. Village ponds provided water for cattle and ducks, and the passing cart in need of a quick wash. The continual coming and going of the animals kept a weed-free area of open water, which was important from the wild animals' point of view, improving the pond's life by diversifying its habitats. Quite apart from the traditional village pond, there were mill-ponds, dew-ponds (set up to serve the distant needs of sheep, and filled by rain, rather than dew) and, indeed, city ponds which served the needs of horse traffic. As time has passed requirements have changed, and most ponds have now been filled to improve the roads. But much wildlife potential has been lost, not to mention character.

Some city ponds remain to give pleasure out of all proportion to the acreage they take up. The pond in St James's Park, London, for instance, is a reminder of the original marsh which was enclosed by Henry VIII to form the oldest royal park. The present design dates from the time of George IV, and it has supported ornamental birds since James I. We are told that Charles II used to make a point of going there to feed the ducks.

## CONSTRUCTING YOUR OWN GARDEN POND

Garden ponds may not have the noble provenance of St James's, but they will give a full measure of pleasure to the owner. It has to be said, however, that there is a certain amount of hard labour involved in the making of even a small one. First of all you must choose the site very carefully: it needs to be level, and must be well away from trees so that it gets plenty of sunlight and does not become clogged with leaves in the Autumn.

For a simple pond which will fit even the smallest garden, the materials required are as follows:

A pond liner, 8ft 3in × 6ft 6in (2.5m × 2m)
Ten wooden stakes
Mallet
Piece of timber, 6ft 6in × 2ft 6in × 2in (2m × 750mm × 50mm)
Spirit-level
Spade
Some sand or old newspaper

The pond liner may be bought from a garden centre, but avoid anything less than 1000 gauge 15⁄1000in (0.375 microns) thick. There are several kinds:

Black polythene – cheapest type, but has a limited life of about five years unless protected from sunlight by a covering of earth and stones. Use only water-resistant type.
PVC sheeting – more expensive but more resistant to sunlight, best type is strengthened with nylon.
Polyolefin – a high grade plastic liner with a life of over fifteen years.
Butyl rubber – a very flexible rubber liner with a life of over fifteen years. The best quality liner but the most expensive.

If you choose to line your pond with concrete it should be at least 6in (150mm) thick and must be sealed with bituminous paint or water-seal cement. But be warned that a concrete pond involves back-breaking work. My advice is to plump for a Butyl liner, even though it is expensive.

## Pond Construction Method

1 Mark out the edges of the pond with stakes; if the ground is not quite level put the shallow end at the bottom of the slope. Hammer in a stake at each end of the pond using the measurements on the diagram. (It's important to keep to these or the liner will not fit.) Rest the wooden beam across the two stakes and mark a point of the beam 2ft (600mm) in from each stake. You can find the position of the other stakes by measuring out from these points.

2 You can now start to dig the pond. Put any turves to one side, as you can use them later to fringe the pond. Save the topsoil as well, because this will be put back in the finished pond. The best of the soil could be used to make a bank. Put soil to one side, out of the way of the spread lining.

3 As soon as the hole begins to grow, check that the sides of the pond are at the same level by laying the wooden beam over the pond and checking with the spirit level. Take out the stakes and remove the earth from the higher sides.

4 Use the diagram to work out the different depths of the pond. The deepest part of the pond is a third of the way from the end, and the sides of the pond should shelve up gradually from this point to the edge. Do not make the slope too steep. Check the depths are right by measuring down from the wooden beam. If you dig too deep don't worry – you can always put some back!

5 A special feature of the pond is the marshy area at the shallow end. Extend it by removing earth to a depth of 2in (50mm) for a distance of 8in (200mm) away from that end.

6 When the hole is ready, pick out any stones or sticks which may puncture the lining and carefully pour in some damp sand to make a layer ¾in (20mm) deep over the whole of the pond. This is to give a soft base for the lining to rest on. If you prefer, you can use sheets of spread-out newspaper, but make sure you build up a good layer.

7 Take the liner and with it overlap the deep end of the pond by 8in (200mm). Anchor it with heavy stones, temporarily.

8 Spread out the liner over the pond, and push it gently down into the hole. Don't try to make it fit too snugly as the weight of the water will do that.

9 Fill the pond almost to the top with water. You can now see whether you've made any mistakes in levelling the pond. Any errors can be put right by removing or adding earth under the liner. Make sure that there are going to be water levels of from ¾in to 4in (20mm to 100mm) at the shallow end.

10 You should wait two or three days to allow the liner to settle, then carefully trim off the surplus plastic round the edge leaving at least 8in (200mm) overlap, but do not cut any off the shallow end. Do not trim too soon, the weight of water drags in a lot of slack. And do not trim too close, or you will have difficulties when arranging your pond surround of stones or turf.

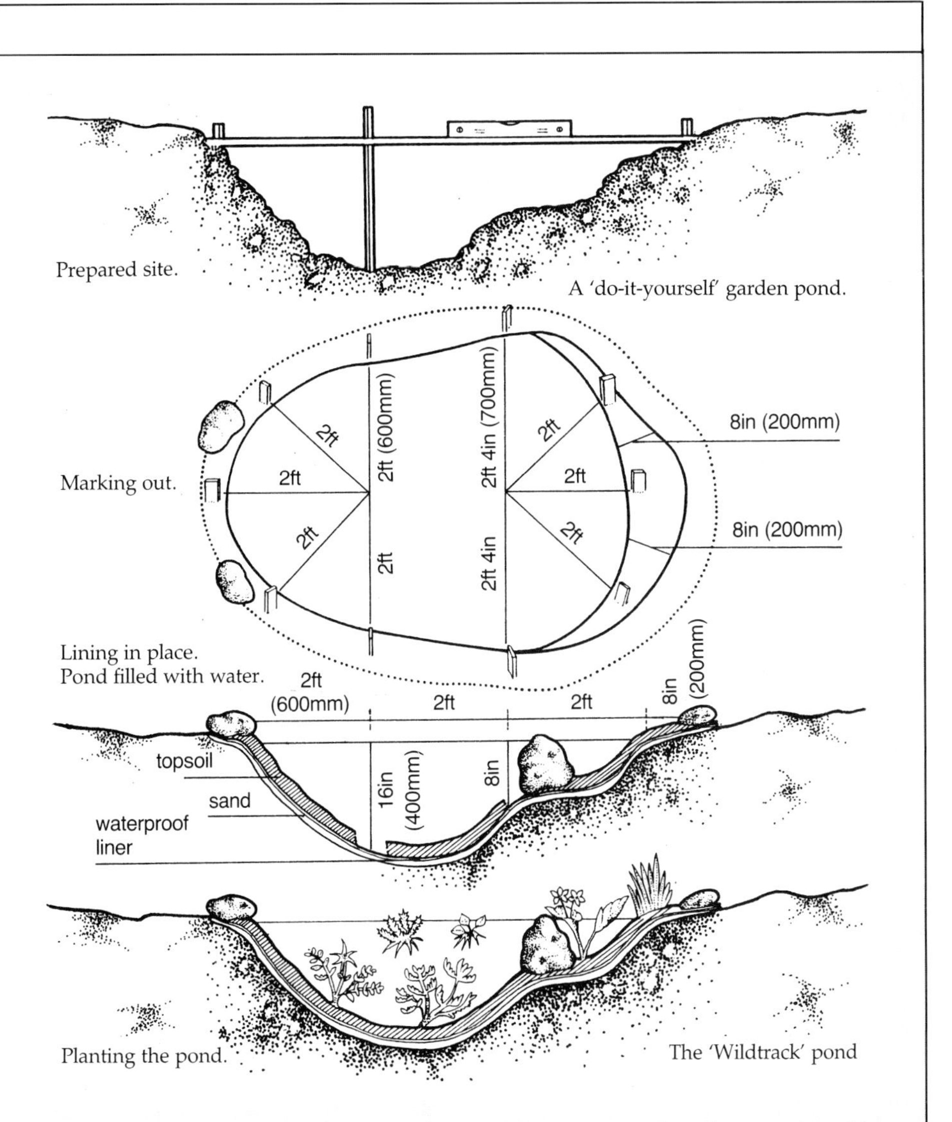


Prepared site.

A 'do-it-yourself' garden pond.

Marking out.

Lining in place.
Pond filled with water.

Planting the pond.

The 'Wildtrack' pond

11 Secure the edges of the plastic by burying it under earth or large stones. You can also use the turf for this purpose. If you are using a polythene liner, it's very important to ensure the sheet is totally buried, as in time sunlight will destroy it.

12 Take the topsoil you have saved and sprinkle it over the surface of the pond until you have built up a layer of silt several inches deep over the liner. The earth will form a marshy area at the shallow end. This earth should not be in contact with the surrounding ground or in dry weather water will be drawn out of the pond.

13 Take a couple of large stones, and carefully place them in the shallow end so that they show above the surface. Birds will use these as perching places.

## POND PLANTS

The pond must be properly stocked with suitable plants for it to establish a healthy home for a living community. The plants release oxygen into the water and absorb the carbon dioxide produced by the water animals which are bound to colonise it. They also provide food, shade and shelter for the animals. They thrive on plenty of light, but welcome protection from strong winds. There are three main plant categories, and you should make certain your pond is stocked with some of each.

1 *Free-floating plants* live at the surface, with their roots suspended in the water, eg Frogbit, *Hydrocharis morus-ranae;* Water Soldier, *Stratiotes aloides;* duckweed, *Lemna* sp; Water violet, *Hottonia palustris* – whose flowers are rich with nectar and attract pollinating insects.

2 *Oxygenators* live fully submerged in the deeper part of the pool, some rooted to the mud at the bottom. They are vital, eg Starwort, *Callitriche autumnalis;* milfoil, *Myriophyllum* sp; Hornwort, *Ceratophyllum demersum.* The Starwort is specially valuable because it retains its oxygenating properties throughout the winter. Quillwort, *Isoetes lacustris,* is an excellent food plant from the point of view of fish.

3 *Marginals* live in marshy areas at the edge of the pond and have most of their foliage above water, eg Water Forget-me-not, *Myosotis* sp; Brooklime, *Veronica beccabunga;* Marsh Marigold, *Caltha palustris;* Flowering Rush, *Butomus umbellatus;* Slender Spike-rush, *Eleocharis acicularis* (this may survive submerged but will then be sterile).

You can find these plants easily enough in wild ponds and ditches, but it is best to buy pest-free stock from a nurseryman. Follow his instructions in planting. But a general rule is to put the marginals in water 1in to 4in (25mm to 100mm) deep, rooting them in good topsoil.

The oxygenators will grow in water 6in to 24in (150mm to 600mm) deep; if they need planting (as *Myriophyllum,* for instance), the best method is to use the specially made baskets which are cheap to buy and allow you to make subsequent gardening changes easily. A cheaper way is to take a piece of lead wire and bend it loosely around the base of the plant before you introduce it to the few inches of topsoil which covers the bottom of your pond.

## POND ANIMALS AND VISITORS

Animals will find their own way to your pond, but it makes sense to introduce some common water snails straightaway, for they will serve a useful purpose in grazing the algae. Buy them from your aquarist, and remember that cheap ones eat just as efficiently as expensive ones. Ramshorn Snails, *Planorbis corneus,* or the Freshwater Winkle, *Paludina vivipara,* are the species least likely to attack your 'best' plants. One snail to every 2sq in of surface water is said to be the desired population, but don't put too many in, they will soon find their own balance.

You may wish to introduce frogs (by way of spawn clouds), toads (spawn strings), water spiders and beetles to suit your own whim. But go easy on the Great Diving Beetle, *Dytiscus marginalis*, if you are going to have largish fish, because it will attack them. And avoid newts in a small pond for they will eat almost anything. Most insect species will find their own way – for instance dragonflies, water boatmen and pond skaters. Sticklebacks and Minnows will control the mosquito and gnat larvae which will inevitably appear. Again, a rough rule of thumb is 1in (25mm) of fish to 24sq in of surface area. On the whole it is probably best not to have any fish in a very small pond.

Birds will enjoy hunting the pond life, you will enjoy the drama. Robins and Blackbirds go for tadpoles, Blackbirds try for newts, Kingfishers enjoy Sticklebacks, and if you are un-British enough to introduce Goldfish or Carp, then you may be fortunate enough to be visited by a heron. But Goldfish are a mixed blessing; they may look colourful but they are bottom feeders, creating in the process a continuous cloud of mud which means the water is rarely completely clear.

Theoretically, the pond should need little maintenance. When it is first filled and planted it will probably be opaque and green-looking for a while, but as the plants grow the water will clear. In hot weather you may well have to add water, especially if you are losing it by capillary action. Devise some way of trickle-feeding it, or diverting rain to it, to save a great deal of trouble. If it is near trees, then fallen leaves may need to be removed in the autumn, but it should not be necessary to clean the pond out. With well-balanced populations a pond will stay healthy for years.

You will find that the finished product will give you as much pleasure as it does the birds, and if, one day, a heron comes and eats your fish you must grin and bear it and re-stock the pond.

It really isn't possible, or natural, to try to run a sort of 'paradise garden' in which predators have no place. However hard it may be for us to reconcile ourselves to it, the fact is that predators like foxes, Sparrowhawks and herons, which prey on and eat other species, are operating in the best interests of the species on which they prey. By catching the slower individuals which are off-colour or sick, they are continually weeding out the less healthy members of a species so that it is the fittest which survive. So please do try to understand both sides of the argument. When the Magpie eats eggs or small birds it is doing its job as a 'magpie', not acting like a pantomime villain!

## LARGER PONDS

A pond which can be comfortably jumped across can provide a great deal of enjoyment as well as serving a useful purpose for the birds. But there's no doubt that the possibilities deriving from a larger water area become a great deal more interesting. If you live within reasonable proximity to a wildfowl flightline, by a

Herons will approve of the fishing prospects in a garden pond.

river valley or water course for example, there is every chance that you will attract breeding or wintering ducks. Even a pool as small as 30sq ft has attracted flighting ducks, but the larger the better. And if your own home patch offers no chances, then why not cast an influential eye on land owned by public corporations or commercial interests? Over the last few years the great potential of gravel pits in south-east England has been realised, very much to the benefit of water birds and migratory wildfowl.

For a large pond or lake one of the requirements is a deep water area – a depth of over 3ft (900mm) – uncluttered by plant growth. This provides for an unimpeded landing place, and also a gathering place for ducklings in the breeding season. Then there must be a shallow area of less than 1ft (300mm) depth which allows for plant growth and duckling cover. An island is most important, since it serves as a relatively safe place for resting and preening, and for nest sites. Obviously there must be a reliable source of clean water available from a stream or spring, or the natural run-off from a few acres of grassland, which is itself important for its feeding potential.

If you are starting from scratch, digging is a better construction method than damming, which is an expensive proposition; and, of course, it is better if you are working in flat, open country. If you must dam then it is prudent to check with your local water authority on the legality of your project, and you should get advice from a professional civil engineer. But it seems that the use of a dragline presents the most practical method of pond making, with the built-in advantage that it makes the provision of a central island very convenient, since the machine simply works its way round a perimeter, clawing out great chunks of soil and leaving the central area untouched. Pond construction is dealt with in some detail in a booklet published by the Game Conservancy (for address see page 183).

If your pond has no island, consider establishing a raft which serves much the same purpose in providing a degree of safety from terrestrial predators like rats and Foxes. Rafts offer the big advantage in that they are immune to changes in water level, and their surface area can be adjusted to suit the size of the pond. They may be constructed of railway sleepers or telegraph poles or even of metal girders on old oil drums. One design involves two 30ft (9m) telegraph poles securely moored and held 1ft (300mm) apart by struts. A 30ft × 18in (9m × 450mm) length of Netlon plastic netting is attached underneath, to support a mat of growing vegetation. This practical device will support nestboxes and also acts as a staging post where birds may loaf about or preen. Organise a ramp which will allow easy access for ducklings. Whatever shape or size of raft you devise, it must be topped with soil, giving a home for vegetation such as *Phragmites* reeds and *Juncus* rushes. In turn, these may attract Reed Buntings and Sedge Warblers to breed. A successful raft will be well used, not only as a breeding place but as a winter roosting area, especially in bad weather. Rafts may be used by divers, grebes, swans, ducks such as Mallards, Tufted Ducks and Goldeneyes, geese such as Greylag and Canada, terns, Moorhens and Coots, offering them safety from predators, shelter from the elements – and a potential house site.

In the case of big ponds and lakes, the associated bankside vegetation can clearly be on a larger scale. When you are dealing with migratory duck, there is a pressing need for seclusion, coupled with shelter from the elements, in the perimeter cover. Waterside cover is well provided by quick-growing birches, and by alder, poplar and the many forms of willow. Some of the most useful willows are listed in the following table.

### Pondside Willows

Creeping Willow, *Salix repens*. Low growing, up to 6ft (1.8m), usually 1ft (300mm), good ground cover.

Common Osier, *S. viminalis*. Low screen, up to 18ft (5.5m), but can be coppiced and kept at 9ft to 12ft (2.7m to 3.6m)

Goat Willow, *S. caprea*. Female known as pussy willow, with silver catkins. Male has yellow catkins. Plant well back from water's edge, seeds and colonises freely, needs control. Achieves 23ft to 26ft (6.9m to 7.8m) in height.

White Willow. *S. alba*. Common in water meadows. Achieves 59ft to 66ft (17.7m to 19.8m) in height. Can be pollarded.

Crack Willow. *S. fragilis*. As White.

Willows take violent pruning, even to ground level, producing dense scrub. They are easily propagated, as anyone who has a willow fence post will know.

An outer belt of sheltering conifers provides a desirable shelter belt, but avoid tall trees close to the pond margin, where they will deny much-needed light to the pond plants, and will also over-enrich the pond with fallen leaves. They also represent obstacles and hazards on the final approach to landing for incoming flights of ducks or swans.

Around the edge of the pond itself, you may want to introduce Wild Celery, *Apium graveolens*, and Millet, *Setaria italica*, as well as the *Phragmites* reed beds and sedges which will act as a screen. For these plantings you will be able to collect your material from local ditches and marshes where they are freely available. Take good balls of root, with emergent shoots, cutting back all the aerial growth to encourage shoots and reduce windage during the period of establishment. Plant early in the growing season, from April to mid July, but the earlier the better. In the deep-water areas, introduce Pondweed, *Potomogeton natans*, and Amphibious Bistort, *Polygonum amphibium*. This latter plant is invaluable because apart from supporting much underwater life in the summer when it is at the surface, it sinks to reveal a clear surface in autumn, when the ducks need an alighting area.

Fence stock well away from the pond, for they will trample the marginal cover. And make provision for foraging areas and nesting areas, the one needing open grassland, the other secret and sheltered brambly places. Both ducks and ducklings need easy access to water and land, by way of a gentle sedge slope.

If you hope to attract wild duck, you should bait the pond with corn. Barley is their favourite, and they will enjoy maize once they become accustomed to it. But any grain mix will do, as well as split peas and beans and even more surprising things like raisins and bananas. Potatoes, especially when frosted and mushy, pulped or mashed, are much enjoyed. Scatter this food, mainly over the shallow edges of the pond, as widely as possible. A two-gallon bucket, which holds a dozen pounds of barley, will feed seventy or eighty ducks, especially if the bottom is gravelly, allowing maximum efficiency in food finding. Feed late in the day, a half-an-hour or so before you expect the birds to arrive. The food should all be eaten by the morning. Do not overfeed. Shooting men, who have pioneered the development of duck ponds, tend only to feed during the shooting season, but there's a lot of sense in starting even as early as June, then continuing till the end of winter, even into April, for the best results.

**Further reading:**

*The Pond Book*, Valerie Porter, Croom Helm, 1989; *Ponds and Lakes for Wildfowl*, Michael Street, Game Conservancy, 1989.

# 4

# NESTBOXES

## HISTORY OF NESTBOXES

### PIGEON HOUSES

Man-made nest sites have a long history, and originally their purpose was almost entirely in connection with the food potential of the species involved. It seems likely that pigeons were the first birds to be domesticated, by way of artificial nesting ledges which encouraged truly wild birds to breed in places where the keeper could harvest his share of the fat squabs. Pigeons were ideal candidates for domestication, with undemanding food requirements, an easily satisfied specification for nesting places, an easy-going disposition and tolerance of man. Above all, they had the astonishing ability to rear as many as ten clutches in a year, nourishing a pair of well-grown squabs while they incubated the next pair of eggs in the production line.

The pigeons' technique for feeding young, even in the depths of winter, relies on their ability to produce 'pigeon milk' from their crops. The formation of this milk is controlled by a hormone, prolactin, which is produced by both cock and hen in the last days of incubation. Thus, they are ready to ensure a protein supply to the newly hatched chicks for the first few days, till they are strong enough to eat at least a little solid food. The milk is secreted from the lining of the parent's crop, which thickens and, at hatching time, breaks off into cheesy curd. The parent takes the newly hatched squab's bill into its mouth, the squab automatically reaches in, and the parent regurgitates. After a few days the milk is supplemented with choice pieces of soft food or seeds. As times goes by the ratio changes, by hormonal cueing, so that the young get less and less milk.

Probably the first of these birds to be domesticated was in the Eastern Mediterranean. There are images of the pigeon in art dating back to 3100BC, and certainly the birds were used as a source of food in Egypt before 2600BC. In early cultures pigeons were sacred birds, associated with Astarte, the goddess of fertility and fruitfulness. In India, as birds sacred to the Hindu religion, they were allowed to colonise buildings and temples without hindrance and without being exploited for food. In Classical Greece they were associated with Aphrodite, as love symbols. The Romans linked pigeons with Venus in the same way, but prudently took advantage of their culinary qualities, as well as using them for messengers. Indeed, as Pliny the Elder (AD23–70) wrote in *Natural History*: 'Many

The Ark. Coloured woodcut from *Biblia Sacra Germanica*, Uton Koburger, Nuremburg, 1483.

persons have quite a mania for pigeons – building towers for them on the top of their roofs, and taking pleasure in relating the pedigree and noble origin of each'.

As enthusiastic pigeon fanciers, the Romans must have built pigeon cities, or *columbaria* as they were known, in England. Whether the Saxons did so is not clear, but the word 'cote' refers in part to a bird house, and 'culver', though less widespread, probably refers to the wood pigeon and gave rise to 'culver house'. 'Doocot' is clear enough in its meaning and, indeed, the Scots have cultured pigeons for a very long time.

Near the coastal village of Wemyss, in Fife, there are two caves with dozens of man-made ledges cut out of the walls and the artificially enlarged roof, clearly designed to encourage Rock Pigeons. Carvings on the walls date back to the early Bronze Age, but are mainly Pictish (AD400–900). Sadly, the archaeologists are as yet unable to date the ledges, so we cannot say whether the Picts farmed the pigeons or whether the practice was first associated with the later development of the nearby fourteenth-century castles of Wemyss and Macduff.

More widespread rearing of semi-domesticated pigeons for fresh meat, especially in winter, began in Britain with the Norman invasion, when the conquerors introduced the science of the 'colombier'. No twelfth-century castle was complete without its rows of pigeon holes, carefully built into a turret or high place, sheltered and south facing, with the enormous advantage that the birds foraged far and wide for their own food and did not require sustenance from a besieged garrison's meagre supplies. Soon the gentle pigeon became associated with warlike places, the substantial stone-built dovecot being an important part of any manor house or monastery.

On the wild Gower coast of South Wales, a natural cliff fissure was enclosed at some time in the thirteenth or fourteenth century to form a hollow pigeon buttress, known as the 'culver hole'. There were ramps giving access for the keeper, and three elegant slits making entrances for the birds – yet another arrangement designed to tempt wild Rock Pigeons to nest so that the occupants of a nearby castle could take a proportion of the squabs. The culver hole was rebuilt in the fifteenth century and still stands today, an astonishing structure which reminds us of the one-time importance of pigeon culture.

By the late thirteenth century a medieval bishop on his travels would expect a high standard of victualling, the dinner table liberally provided with bird meat. On tour, his chaplain would record disbursements for wine, beer, beef and so on. One note tells of His Lordship's purchase of '2 carcasses of beef, 9s 4d, 25 geese 5s 2½d, 24 pigeons 8d'. So along with the fish pond and the rabbit warren and the duck decoy, the dovecot was an important item in medieval animal architecture. Conradus Heresbachius, in his *Husbandrie*, 1577, wrote 'it behoveth especially to have good care for breeding of pigeons, as well for the great commoditie they yeelde to the kitchin, as for the profit and yearely revenue that they yeelde (if there be good store of corne seedes) in the market'.

The pigeon house was always carefully sited, to provide protection from prevailing winds. The Norman design involved a circular building, solidly planted on the ground with walls 3ft (1m) thick (no windows), gradually tapering, at the very top of the roof, to a 'lantern', which gave entrance for the birds. A single door at ground level allowed entrance for the pigeon keeper. Inside, the walls were covered with row after row of pigeon-holes, 'three handfulles in length, and ledged from hole to hole for them to walke upon'. An ingenious device called a potence allowed the keeper to reach any nest by means of a ladder which rotated on a central pillar passing within a couple of inches of the wall as it was pushed around. The interior was dim, to the liking of the pigeons, and each pair of birds had a double nest site because before one set of squabs was ready to leave the nest the hen might well have laid her next clutch of eggs.

As time went by the pigeon houses improved in design and, especially in areas lacking local stone, they would be built of timber or brick and set on pillars. This had the great benefit of affording protection from predators such as rats, cats, Weasels and squirrels. Hawks and owls were also a problem, and entrance holes, however carefully placed, attracted unwanted visitors. Heresbachius wrote in 1577: 'I found of late in myne own Dovehouse, an Owle sitting solemnly in the Nest upon her Egges in the middest of all the Pigions, by reason of the thickness of his feathers, yet will creep in at as little a place as the Pigion will; so small and little is their bodies, though they be bombased with feathers'.

Up on the roof, the pigeons would have a promenading area, sheltered from cold winds and facing south to catch the best of the sun. Under the whole structure there might well be room for a stable or cow house, giving the benefit of extra warmth in winter, to encourage breeding. Everything was done to aid these 'wondrous fruitfull' birds. Each year the best of the early squabs were carefully selected for breeding. The less satisfactory – 'unfruitefull and naughtie coloured, and otherwise faulty' – quickly found themselves being fattened for the table. But the right to husband pigeons was a privilege enjoyed by the chosen few – the nobility, the lord of the manor and the clergy – those who were powerful enough to be able to lay their hands on just a little spare corn in the winter. The peasants, and doubtless others as well, made do with clay pots set up to attract sparrows and Starlings, from which they took the first broods of nestlings when they were fat enough to eat. This procedure, using wooden cistulae (flasks), was used in Silesia; and practised in Holland, with unglazed earthenware pots, in the late Middle Ages. In France they hung similar earthenware pots under the eaves of the houses in the region of Toulouse. When the Dutchmen came to drain the East Anglian fenlands in the mid-seventeenth century they brought the practice with them. Before this time sparrows had been paid as part of the rent, according to Norfolk rent books dating back to 1533, so one assumes the locals already had some knowledge of the technique.

For those whose status ran to it, pigeons offered a far better return than lowly sparrows. By the end of the seventeenth century, John Smyth could write of the Berkeley family of Gloucestershire: 'In each manor and almost upon each farm house he had a pigeon house, and in divers manors two. And in Hame and a few other (where his dwelling houses were) three: from each house he drew yearly great numbers. As 1300, 1200, 1000, 850, 700, 650 from an house. And from Hame one year 2151 young pigeons.' These squabs must have represented an important source of revenue, fetching 2d a dozen, a worthwhile sum at that time.

Not only the fat squabs had value: the plentiful droppings which piled conveniently on the dovecot floor were rich in nitrogen and minerals. 'Doves dung is best of all others for Plants and Seeds, and may be scattered when anything is sown together with the seed, or at any time afterwards. One basketful therof is worth a cartload of sheep's dung. Our countrymen also are wont to sow Doves dung together with their grain' (Francis Willughby, *The Ornithology*, 1678). In Persia the dung was used to fertilise melon fields; in south-west France, where it was used in vineyards, dovecots were often kept as much for the value of the dung as the squabs.

Dung was also used in the tannery process, in removing hair from the hides, and as part of the process in making saltpetre for gunpowder. In addition it was used as a specific against the plague and the palsy. 'The flesh of young pigeons is restorative and useful to recruit the strength of such as are getting up or newly recovered from some great sickness' (Willughby, as above).

But the privileged aspects of pigeon-keeping involved social injustice. While they might be owned by the lord of the manor, the birds thought nothing of eating the peasant's or yeoman's corn. And lawyers offered no redress, as the jurist, John Selden, wrote in the early seventeenth century: 'Some men may make it a case of conscience whether a man may have a pigeon house, because his pigeons eat other folk's corn. But there is no such thing as conscience in the business: the matter is, whether he be a man of such quality that the state allows him to have a dove house; if so, there is an end of the business: his pigeons have a right to eat where they please themselves.' So the yeomen and tenant farmers gradually tore the shaky fabric of this law to pieces, building their own version of the free-standing pigeon houses, opening neat rows of nest-holes along the walls of their barns and farmhouses. But in any case the centuries of the pigeon house were coming to an end, as new agricultural practices made it possible to sustain more cattle and sheep through the winter months, thus making butcher's meat available throughout the year.

Most pigeon houses now lie derelict, while their inhabitants' descendants make a living in our city streets. Their truly wild ancestors have fared little better. Sadly, the wild coastal Rock Pigeon is becoming increasingly rare. A number of small populations of the pure-bred birds exist, confined to the north and west coasts, and Scottish and Irish islands. These are bluish-grey birds, with two striking black bars across the folded wing, and with neck and breast an iridescent purple and green. Apart from those far flung Celtic outposts of pure-bred birds, Rock Pigeons are now represented only by the cliffside colonies of racing pigeons which have given up the sport, and by the ubiquitous street pigeon which itself represents a hopeless mix of long-ago-escaped dovecot stock and more recent racing pigeon dropouts. Mixed or not, all domestic pigeons – racers, dovecot, messenger and fancy breeds – owe common ancestry to the Rock Pigeon, with its built-in suitability for intensive breeding in nestbox pigeon holes.

## Wildfowl

Apart from pigeons, wildfowl have provided the earliest examples of artificial nesting devices. Various duck have been encouraged to nest in places convenient for their harvesting. Goldeneyes, for instance, nest naturally in tree holes or tree stumps, and by the late Middle Ages the Lapps were improving natural sites to attract them as a food source, mainly for the eggs.

Not long after the Viking colonisation of Iceland, coastal farmers realised the special qualities of the breast feathers of Eiders. Lining the nest with its breast feathers, this sea duck arranges an eider-down quilt to cover and retain the warmth of the eggs if she leaves the nest for any reason. By the cunning provision of carefully placed sticks and stones (the birds like to nest against something), the farmers created conditions which suited the ducks and so encouraged the formation of colonies in places which were convenient for the down collector. The practice is still followed today, and the farmers go to great lengths to please their worker birds, which are fortunately very tame. They provide music in the form of wind-activated instruments and hang coloured ribbons in string, both of which are supposed to act as added attractions. Some of the Eider colonies are large, with anything up to ten thousand pairs nesting. The down is taken twice in each season, once just before the eggs hatch, when the lining is removed, and again after the young have left the nest. The down, which is carefully cleansed of any dirt and grass or large feathers, represents a substantial income to the farmers. They have a vested interest, of course, in making sure the wild duck breeds successfully and continues to patronise the facilities so carefully provided.

Eiders are one of the most numerous duck species in the world, their winter population in Europe totalling more than two million. Since the mid-nineteenth century their range has been expanding and we now have sizeable breeding

numbers in Scotland – perhaps one day we shall see the birth of a new British industry. In passing, we should note that the same basic technique pioneered for attracting Eiders to nest, that is the provision of suitably shaped sticks or stones in a featureless landscape, has been used by egg collectors wanting to plunder the eggs of Greenshanks, whose moorland nests are notoriously difficult to find. (Nowadays it is illegal to take any eggs.)

Doubtless Mallard ducks, and others, have been provided with convenient nest sites (convenient for the plunderer, that is) in Britain for centuries, but it is not easy to find evidence of them. Decoys, which date back to somewhere round the thirteenth century, were set to live-trap migrant wildfowl in winter, and this practice certainly led to a small-scale provision of nesting facilities in the breeding season. Perhaps the first record of duck boxes in Britain comes from the diarist John Evelyn, a friend of Samuel Pepys and a mine of information on the times of Charles II. In his diary entry for 9 February 1665, he writes: 'I went to St James's Park ... at this time stored with numerous flocks of ... wildfowl, breeding about the Decoy, which for being so near so great a city, and among such a concourse of soldiers and people, is a singular and diverting thing... There were withy-potts or nests for wildfowl to lay their eggs in, a little above the surface of the water.' The designer of the decoy, incidentally, was a Dutchman who had been brought over especially to do the job, Sydrach Hileus.

In North America, the Indians used nestboxes made from bottle gourds to attract birds to nest and offer themselves as a food source. The disadvantage of this method is that the gourd had to be smashed to get at the contents. But they also, more effectively, farmed their wildfowl. Canada Geese, relatively tame and easy going in their choice of nest sites, were obvious targets. Over the years they were greatly exploited by men who found that they took freely to haystacks and sheltered positions in the lee of a fence or hut. Introduced to Britain as a status symbol to decorate the fashionable landscaped lake in the eighteenth century, they bred freely on the islands so conveniently provided for them. Indeed, they flourished to become a common feral goose breeding all over England, a fair part of Wales and with outposts in Scotland. They now enjoy popular protection in city parks and places like gravel pits and semi-natural reservoirs.

Black-headed Gulls have been farmed for their eggs in times past. Breeding colonially, they patronise lake islands, sand and shingle banks at the coast and inland. From the eighteenth century onwards traditional gulleries were further encouraged by the preparation of man-made islands called hafts. In midwinter great quantities of reeds and rushes were cut and level places laid out to greet the return of the breeding birds in spring. These gulls lay in April, and the pairs were so numerous that at Scoulton Mere, East Dereham, in Norfolk, for instance, some thirty thousand eggs were taken annually, with forty-four thousand marking one particularly successful year. At Pallinsburn Hall, in Northumberland, a seven-acre

*Above left:* The well-preserved dovecote at Felbrigg Hall, Norfolk. (National Trust). *Tony Soper*

*Below left:* The 'potence' was an ingenious ladder device which allowed the pigeon keeper access to every one of as many as five hundred pigeon holes. *Tony Soper*

lake was said to be 'covered so thick, when they are disturbed and on the wing, as if a shower of snow were falling on it'! (*History of British Birds*, F. O. Morris, 1857.) The young birds were also considered good eating, some gullery proprietors making fifty to eighty pounds a year by their sale. At the beginning of June, when the young were near fledging, they were driven onto the bank and netted, 'an occasion for jolity and Gentry'. In Staffordshire, three days of netting, over a period of a fortnight, gathered fifty dozen fat gull chicks at each drive.

## NESTBOXES FOR PLEASURE

So far, all these examples of artificial nestboxes or man-assisted nesting have had a culinary or commercial significance. Perhaps the first known record of birds attracted to an artificial nest site purely for aesthetic reasons, was that of Gilbert White's brother Thomas, who, as Gilbert recorded in his journal for 5 June 1782, 'nailed up several large scallop shells under the eaves of his house at South Lambeth, to see if the House Martins would build in them. These conveniences had not been fixed half an hour before several pairs settled upon them; and expressing great complacency began to build immediately. The shells were nailed on horizontally with the hollow side upwards; and should, I think, have a hole drilled in their bottoms to let off moisture from driving rains.'

Charles Waterton, whose pioneer wildlife reserve was described in chapter two, developed the use of nestboxes in the early nineteenth century, as also did the other pioneer field ornithologist, J. F. Dovaston. Unfortunately, although we know Dovaston used boxes in connection with his experiments, which may have been the first to consider the principles of territory in bird behaviour, he left precious little published information on them – a terrible lesson for all amateur scientists who fail to record their findings! Waterton wrote fully, so we know for example that he made (possibly in 1816) an artificial sand quarry with fifty deep holes in a sheltered and sunny part of the grounds of Walton Park. And we can imagine his pleasure and delight when, the very next summer, Sand Martins arrived in his reserve for the first time to found a thriving colony. (A similar experiment, equally successful, may be seen today at the RSPB's Minsmere reserve in Suffolk.)

Waterton improved hollow trees to make them more attractive for Tawny Owls, and developed a Barn Owl house 'on the ruin of the old gateway, against which, tradition says, the waves of the lake have dashed for the best part of a thousand years. I made a place with stone and mortar, about four feet square, and fixed a thick oaken stick firmly into it. In about a month or so after it was finished, a pair of Barn Owls came and took up their abode in it. I threatened to strangle the keeper if ever, after this, he molested either the old birds or their young ones.' Waterton was so delighted with his success that he subsequently built four other owl establishments, all of which were occupied. He also built a tower for Jackdaws, rather in the style of a garden dovecot. Its stone pillar, smooth and vertical, was surmounted by a flat circular stone with sharply sloping edges, measures all designed to discourage rats. On top of this he placed a circular stone

Waterton's owl house.

house, with conical roof, each course of stones having some loose ones, channelled to allow inspection access to the nest chamber behind. Although Waterton was regarded as little more than an eccentric in his own time, many of his ideas sowed the seeds of a whole new attitude to wildlife which were to bear a great deal of fruit later.

By 1897, twenty species were known to have bred in boxes or platforms of some kind in Britain. But the pioneer of large-scale bird manipulation by the use of nestboxes was the Baron von Berlepsch. His primary interest was the control of

The 'old gateway', Walton Hall.

insect pests in his woodland, but there is no doubt he had aesthetic considerations firmly in mind. His main interest was in methods of increasing woodland bird populations in areas where foresters were intolerant of trees past their maturity, and nestboxes played an important part in his operations. Before his time, these devices had been relatively ineffective. He brought a cold and logical eye to the requirements and pursued them with relentless efficiency, pouring scorn on bird 'inventions' which had suffered failures in the past, not being based on what he saw as an understanding of bird nature.

Much of his experimentation, in the 500-acre bird park set aside for the purpose, lay in attempts to design the perfect woodpecker nestbox, having observed that woodpecker nest-holes, deserted or uninhabited, were preferred nest sites by many other species. He proposed that his all-purpose box, whether destined for tits, Nuthatches, Starlings or indeed woodpeckers, was to resemble

A Robin feeding its young in an open nestbox.

The Baron von Berlepsch cut down trees with woodpecker hole-nests in order to measure them precisely.

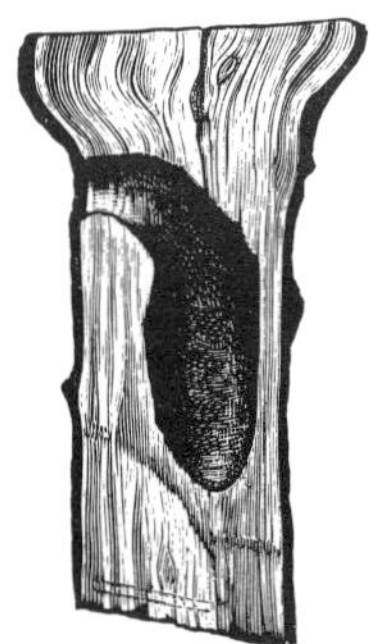

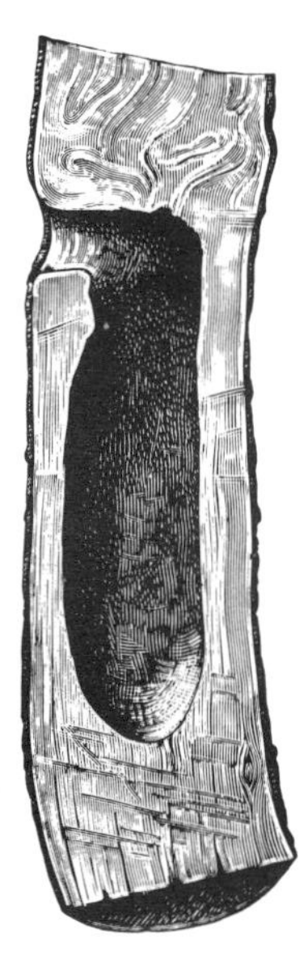
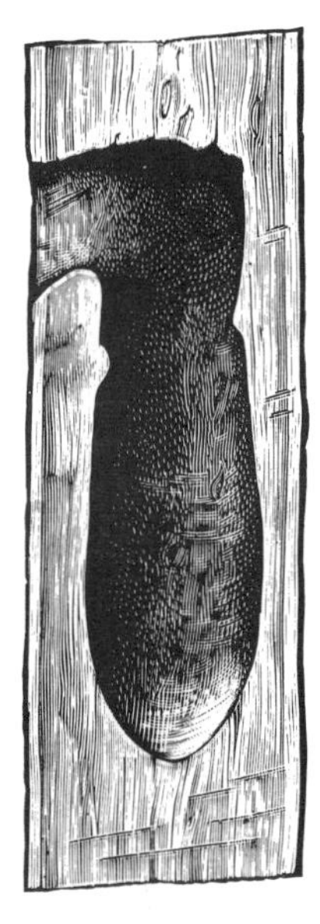

Robins are happy to share your garden, especially if you provide them with a home: a kettle makes a good nestbox, but fix it firmly in a sheltered position and make sure the spout points downwards to act as a drainpipe. *John Clements*

the natural woodpecker design in every exact respect. He proposed no mere invention, but exact copies of nature. He cut down several hundred trees in his search to reveal the woodpecker's secrets, discovering, to his surprise, that the nest cavities were all constructed to exactly the same general principles whether the carpenter was a Black, Green or Spotted Woodpecker. He then set about reproducing, in quantity, the perfect nestbox.

In fact, Berlepsch went too far, since as we know only too well, birds will occupy boxes of almost any shape, size or colour provided they offer certain fundamental design advantages – most particularly that the entrance hole is of the correct size. But, after measuring hundreds of natural woodpecker excavations, he specified that his boxes should reproduce those measurements precisely. He wouldn't allow tin guards around the entrance holes (to discourage Great Spotted Woodpeckers from taking over from tits) on the grounds that they destroyed the natural appearance of the boxes ('their chief merit'), and said that such 'guarded' boxes were never occupied, a claim which seems nonsensical today.

In one wood he set up 2,000 of his boxes, and claimed ninety per cent occupancy; and in his bird park, he had 300 boxes occupied by birds of 14 species. But although he went to some length to provide additional food for his birds in order to sustain an abundant population, it is not clear that he fully understood the overriding importance that food availability has in controlling bird numbers. He believed that the provision of nest sites was of paramount importance. Nevertheless, he was the first to make nestboxing popular; his boxes were successful, they were manufactured in large quantities both in Germany and – under licence – abroad. In the early years of this century the newly formed RSPB offered them in various sizes, for sale at prices varying from 1s 6d (7½p) to 5s 6d (27½p). Berlepsch left his mark, if nowhere else, on the commercial world of nestbox production, and it was years before cheaper, less natural, but equally effective, versions were on the market. In one major respect Berlepsch strayed off the path of righteousness: he believed firmly in controlling predators, creatures he

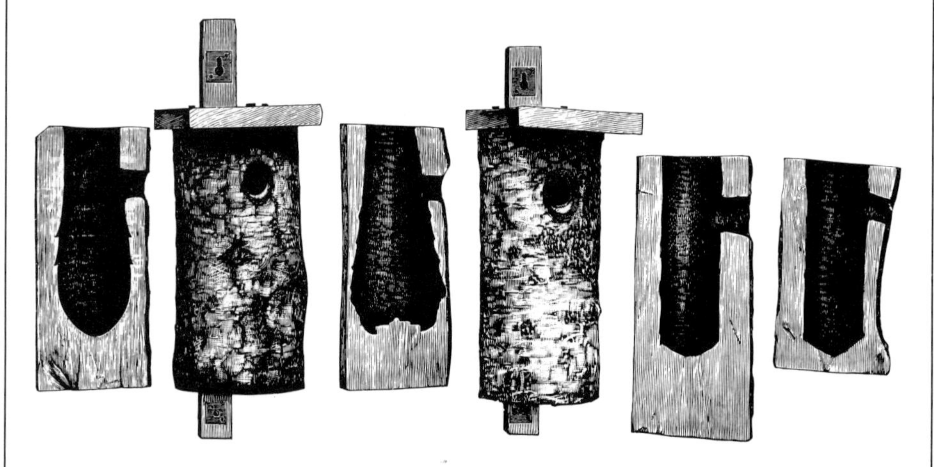

Berlepsch designed his nestboxes to reproduce a woodpecker's nest hole exactly, believing that this was the preference of many other species.

Berlepsch nestboxes were sold at the turn of the century by the infant RSPB.

called the enemies of his birds. These included cats, squirrels, Weasels, martens, Polecats, House and Tree Sparrows, shrikes, Sparrowhawks, Goshawks, Jays, Magpies and Carrion Crows – his blacklist for whose demise he offered rewards.

From Berlepsch's time, boxes were used systematically by scientists engaged in population studies. In the case of the Pied Flycatcher, which takes to nestboxes as ducks take to water, whole woodland populations have preferred the artificial sites to anything a tree has to offer, and tens of thousands of birds have been ringed in research directed towards analysing their life-style. A great deal of work has also been done on the nesting behaviour of Blue and Great Tits, and of Tawny Owls.

## IMPROVEMENT OF NATURAL NEST SITES

While bird table food is at best a poor substitute for natural food, man-assisted nest sites are very often a distinct improvement on the natural variety. Birds manage perfectly well without our help, of course, but the direct pruning of an awkward twig here and the enlargement of a hole there will often make a marginal nest site a prime one. A large, mature garden attached to a well-worn house will offer many desirable building sites to a house-hunting pair of birds. Standard trees and fruit trees provide strong foundations amongst their branches and, as they decay, cracks and openings allow access to secret cavities. Hedges and stone walls, old sheds and rickety eaves all allow birds to prospect and occupy likely spots. No possibility will be overlooked. Even a small garden and an over-maintained house will be colonised, especially if there is a nearby park, or some woodland, to improve food potential. From a bird's point of view, a newly constructed house with a raw garden is the least attractive prospect. For the bird gardener, he has all the enjoyment and anticipation of years of conscious construction and manipulation.

Crudely divided, there are birds which nest in holes and those which don't. Tits, Nuthatches, Starlings, Tree Sparrows and woodpeckers, for instance, live in secret caverns and crevices; Robins, Blackbirds and Spotted Flycatchers live in the

SITE IMPROVEMENT

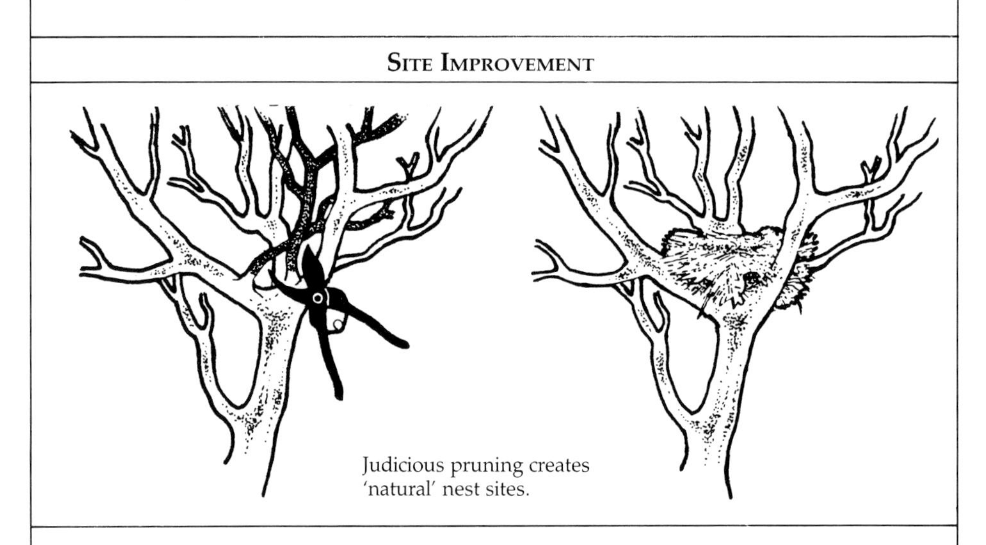

Judicious pruning creates 'natural' nest sites.

open, but for all that they are well concealed. Rooks and herons don't bother to conceal their nests, and instead site them comfortably high off the ground. Most seabirds, waders and waterfowl prefer wide open spaces, yet carefully choose places where they will be undisturbed for other reasons – high cliffs, remote places or islands.

Most woodland is managed for maximum timber production and decaying trees are not tolerated. So there is a chance for the bird gardener to redress the balance if he is able to allow an old and dying tree to live out its time undisturbed. It will give interest out of all proportion to its cash value.

It may be that you already have some old trees – fruit trees, perhaps – which have begun to decay in a manner attractive to birds. If not, you might consider introducing some holes into a decaying tree with a brace and bit. Start some promising holes of about 1¼in (32mm) diameter and a woodpecker may finish the job. If the woodpecker gives up, a Nuthatch may take over and plaster the entrance hole with mud to suit its own preference. Or, you might try importing an old tree or tree trunk, complete with holes, and setting it up in a secluded position. At worst, you will end up with a pair of Starlings. Personally, I like to have a pair of Starlings nesting nearby because they are such entertaining vocalists and are remarkable mimics. We once had one that gave a firstclass rendering of a hen that had just laid an egg, but the climax of its repertoire was a beautiful pussy-cat's miaow.

The famous film maker Eric Ashby spent a rewarding couple of seasons filming woodpeckers which had nested in a rotten birch. When they finally abandoned the tree he carefully sawed it down and took it home to set up again in a likely spot. If you do drill holes, or erect a tree-trunk, try to ensure a bit of shelter for the

**WALL IMPROVEMENT**

This cavity wall mounting nestbox is designed to fit the wall of any new building constructed of standard sized house bricks. *EcoSchemes Ltd* (see page 186).

entrance and face it away from the hottest sun, which can exhaust nestlings.

Hedges offer a multitude of prime building sites for birds, especially when they are prickly enough to discourage predators. Hawthorn and Holly are both excellent. Layered and trimmed, they provide dense cover for Robins, Dunnock, Wrens, Linnets, Greenfinches and Chaffinches, Whitethroats and Blackbirds, and so on. Beech and yew hedges also serve well, and although they are less prickly, they are still difficult to penetrate. Very often you can improve their potential by judicious pruning of forked sites at about 5ft (1.5m) above ground level. Prune in autumn or winter to avoid disturbance at breeding time. Hazel needs special treatment, since it branches from the ground and lacks the forking structures which provide nest support. A little ingenuity is needed to devise some framework at a suitable height.

When you have finished working on trees, you might turn your attentions to the walls of your house and outbuildings. Quite often it is possible to enlarge cracks so that there is the 1⅛in (29mm) necessary for a tit to squeeze through. I am not suggesting you tear your house apart just for a few birds, but you will find plenty of likely and safe places if you look around, armed with a strong auger or jemmy. Try drilling some discreet 1¼in (32mm) holes in your garage doors. Make one entrance up near the roof and Swallows may colonise the loft, although you will need to have a ceiling to protect the cellulose of your car from the droppings.

If you are building a garden wall, do not overdo the pointing; leave one or two gaps and you may attract a Pied Wagtail. Even Grey Wagtails take freely to man-provided nest sites in culverts and bridge stonework. The height of the holes and cavities is not vitally important, although round about the 5ft (1.5m) mark is probably ideal. With a desirable site, birds will not be too choosy. Robins have nested at ground level and Great Tits as high as 24ft (7.2m), although these are exceptional instances. The holes should, however, be in a sheltered position and facing somewhere within an arc drawn from north through east to south-east. Hot sun is bad, and so is an entrance facing into a cold wind. Extremes of temperature can easily kill young nestlings.

An ancient, disintegrating stone wall is an asset to cherish and so is an old garden shed. The wall may be a haven for tits, Nuthatches and wagtails, and the shed may be a thriving bird slum in no time at all if you develop it a little. Shelves around the walls and under the roof at different heights could provide homes for Swallows, Blackbirds and Robins. A bundle of pea sticks in a corner may make a home for a Wren. Leave an old tweed coat hanging up with a wide pocket gaping open for a Robin. Keep the floor clear, though, to discourage rats. If necessary, put a rat-trap tunnel against the walls, but see that it does not let in light and attract ground birds. Make sure there is a good entrance hole somewhere, in case a bird is locked in by mistake.

On the outside of the shed grow a jungle of creeping Ivy and Honeysuckle, for it may well entice a Robin to build. Hide a half-coconut (with a drain-hole in the bottom of the cup) in the creeper for a possible Spotted Flycatcher. Try excavating a nest cavity in the middle of a brushwood bundle and lean it against an outside wall. Lastly, lean an old plank against the dampest, darkest wall to make a haven for snails, and farm them on behalf of the thrushes.

## MAN-MADE NEST SITES

The traditional back-garden nestbox is the Robin's kettle, stuck 5ft (1.5m) up in the fork of a tree. The kettle should be at least quart size, and the spout should point down so that rainwater can drain away. As an encouragement, prime the nest with some dead leaves or a plaited circle of straw.

The most unusual man-made nest site I ever saw was a birdcage, hanging high up on a cob wall in the village of Middle Wallop in Hampshire. It turned out that a Hoopoe had reared young in a perfectly normal cavity-site in the wall, but one of the nestlings had fallen out of the hole a bit too soon. The villagers had hoisted up the birdcage and left the door open, installing the unfortunate young bird inside. Incredibly, the adult Hoopoe carried on feeding the baby as if nothing had happened, going in and out of the birdcage as if it were the most ordinary thing in the world. The story has a happy ending too, because the whole family finally fledged successfully and flew off. But Hoopoes are rare breeders in this country, so I am not going to suggest that we all invest in Hoopoe birdcages.

Every year for some time past a pair of Mute Swans has tried to nest on a tidal sandbank near the mouth of the estuary at Newton Ferrers in South Devon. They seem unable to grasp the fact of tidal movement, and as fast as they build a nest the rising tide washes it away. The scientists of the International Paint Company, who have a research station nearby, came to the rescue, constructing a special swan raft on empty oil drums. They anchored it in position and piled the beginning of a swan's nest onto it. The swans took it over and now every year they complete their nest, lay eggs, brood and hatch the cygnets; every time the tide comes in, the nest and contents rise gently and float, serene and safe.

Rafts have also been used successfully to encourage swans, Greylag and Canada Geese, ducks, grebes, terns, Moorhens and Coots. Their great advantage is that they give a degree of safety from land-borne predators, but they are easier to write about than to construct. They consist of a platform supported by buoyancy tanks, drums or steel tanks, held by a framework of timber or angle iron. The platform, or deck, should be devised to carry a layer of soil or shingle stabilised with suitable plants; polythene sheeting will help to retain sufficient rain water to keep the soil wet. Make sure that there is a ramp or some sort of suitable access point so that the birds can launch themselves into the water and get back onto the raft easily.

In the Dee estuary, Common Terns successfully colonised a raft moored on a reservoir belonging to the British Steel Corporation. The raft, constructed by the Merseyside Ringing Group, was a massive affair of telegraph poles decked with railway sleepers, covered with slag, shingle and grass sods. Blocks of expanded polystyrene provided the buoyancy. Secured by nylon lines to scrap-iron anchors, this incongruous device floated serenely in a scene characterised by slag tips, blast furnaces, power stations, and the constant disturbance of a modern steel works. Yet the terns arriving in spring from their West African wintering grounds took up residence and successfully defended their new territory against the attentions of Herring Gulls.

Mute Swans nesting on a raft.

As the months went by a colony of terns became established, and in that first year the Merseyside group ringed sixteen young birds. In this way, what had been a steady decline in numbers at the nearby old-established ternery was halted, and the Dee tern colony was much strengthened.

All over Europe there are traditions which have led people to encourage birds to adopt man-made nest sites. In many villages in Germany, a cartwheel is fixed to the top of a high pole in the hope of attracting a stork to nest. And in Switzerland and Holland they have developed a successful pole-top nestbox for Kestrels. If you have a large secluded garden in good Kestrel country (and this includes cities), you might consider the experiment of putting up one of these boxes. Farmers who are bothered with mice might also think it worth the effort. (For constructional details see page 110.) Kestrels are among those species which have learned to take advantage of the urban life, taking over window-boxes high up on tower blocks. Nowadays Kestrels breed freely in towns, and even Peregrines are taking to the city life.

There are two basic types of nestbox: an enclosed space with a small entrance hole, and a tray or ledge with or without sides and roof. The closed type tends to be the most successful, partly because the average house and garden don't offer many convenient cavities reached by a suitable tunnel, and mainly because most of the birds which are prepared to live close to us derive from a woodland habitat.

Nestboxes are readily available from commercial suppliers (see the advertisements in the RSPB's magazine *Birds*, and in the monthly magazines *Birdwatching* and *British Birds*, and the addresses on pages 184–5). But beware of novelty designs which owe more to sentiment than biological requirement, ie boxes whose dimensions are incorrectly formulated or which offer footholds to predators in the shape of ornamental windows and chimneys and, in some cases, unnecessary perches. Rustic boxes, made of hollowed-out birch branches, are perfectly satisfactory, although the wood deteriorates rapidly and they don't last many seasons. But there seems little point in going to all the trouble of making

BASIC NESTBOXES

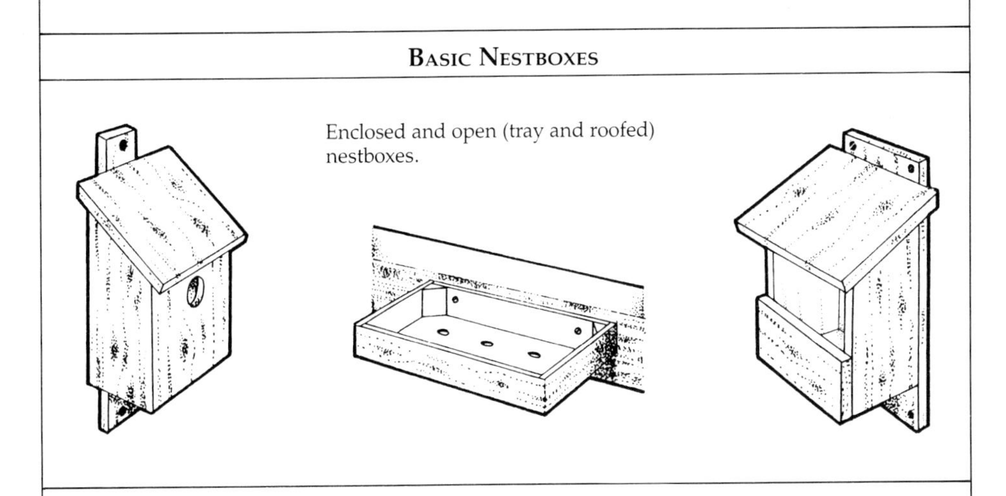

Enclosed and open (tray and roofed) nestboxes.

them when they are usually fixed to almost any tree but a birch, thus losing the possible advantage of camouflage. Plain square boxes are the easiest to make, and there is precious little evidence that the birds mind one way or another. The interior size, entrance hole size and the placing of the box are what count most.

Since tit boxes are most people's introduction to the nestboxing craft it might be most useful to consider their construction in some detail. The drawing (overleaf) shows how to mark up a piece of boarding, 41in × 6in (1040mm × 150mm), to cut out the pieces for one standard box. The ¾in (18mm) timber is thick enough to afford good insulation and to last a reasonable length of time. Softwood such as old floorboarding is usually the right size and is well seasoned and ideal. Cedar weathers well and needs little maintenance.

Hardwood will prove more weather resistant – oak is best – but is more difficult to work with and is prone to splitting, quite apart from being expensive (unless you can get offcuts from a mill). Seasoned material has the great advantage over green or fresh-cut wood in that the latter is likely to warp and split as it dries. The interior surface need not be planed, as a roughened one offers a grip to the chicks as they come to scramble out of the nest.

The insulation properties of the material are worth consideration. It has been shown that a warmer nestbox encourages earlier laying by Great Tits, for instance, and thereby increases the chances of success for the clutch. The laying date of the hen is influenced by her ability to feed sufficiently well to make an egg every day, as well as nourish herself. So, if the bird roosts in a relatively warm nestbox before the breeding period, she is in laying condition at the earliest possible time.

## THE CLASSIC TIT BOX

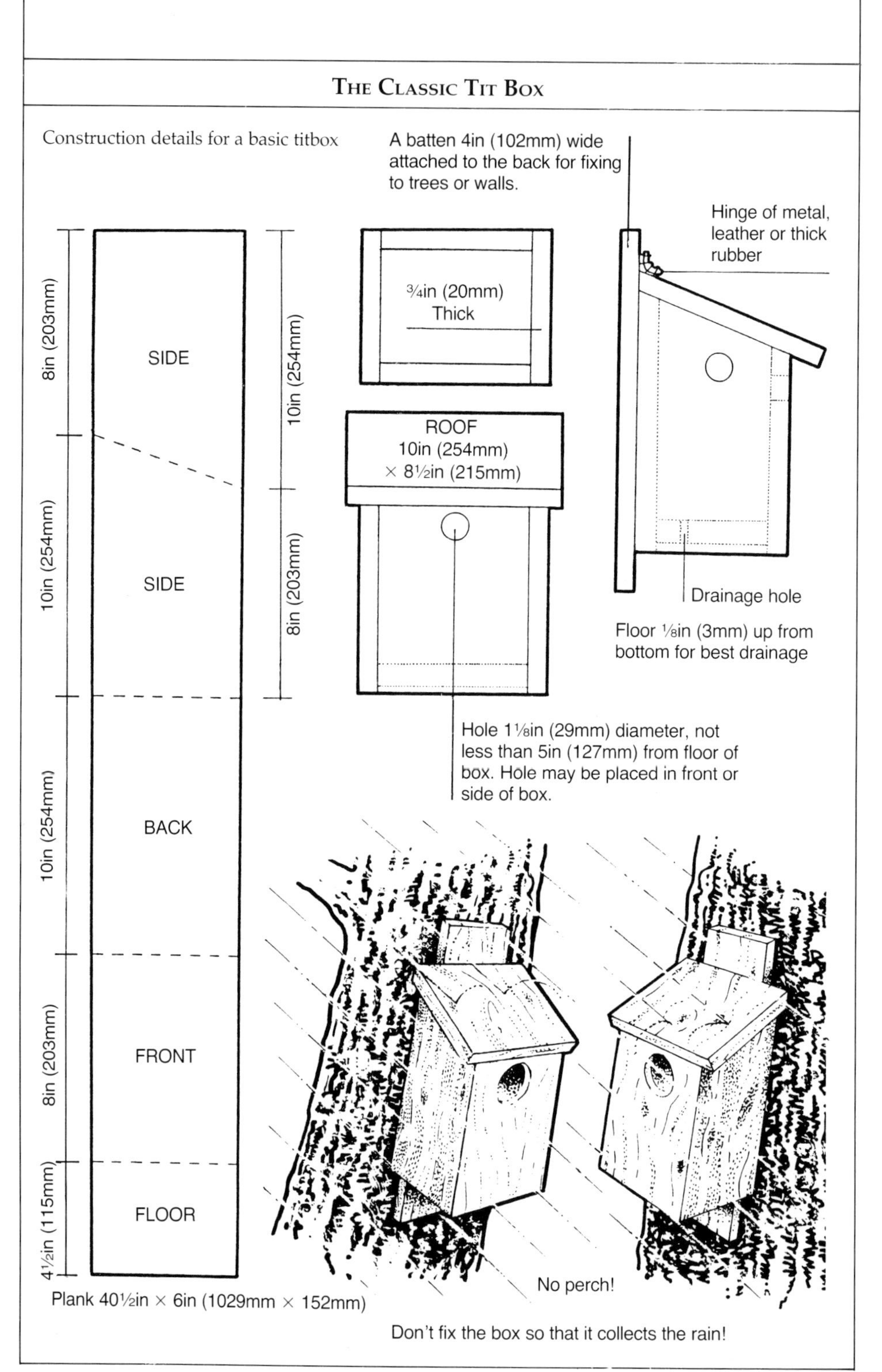

Plastic materials suffer from the bugbear of internal condensation and the danger of chick loss from damp conditions. Concrete has the disadvantage of being heavy and difficult to fix in place. A felt roof may be stripped by squirrels, a thatched roof may be plundered by sparrows. On the whole, timber is probably the most convenient nestbox material.

The interior size is critical. In the case of Great and Blue Tits the floor area should be at least 4in × 4in (100mm × 100mm). In our plan we have allowed for a more generous 6in × 4½in (150mm × 115mm). If the floor is larger than this it merely forces the tits into importing unnecessarily large quantities of material to form foundations for the nest cup. You may feel the dimensions of the interior of the box are surprisingly small, but the incubating birds adopt a squatting position which uses little space, and there is no need to accommodate their stretched length. Also, when the chicks are hatched they will benefit from the warmth of a jumble of bodies, provided there is enough room for them to stretch their wings.

Allow the sides to extend a short distance below the floor, so that draining water rots the wall bottoms before it begins to rot the floor section. The roof or an upper wall section should be removable, so that you can inspect the contents and carry out routine maintenance. The standard RSPB tit box (see page 87) has a front section which lifts out, and the advantage of this design is that there is less likely to be trouble with a leaking roof, but it does need to fit snugly and securely when in position. There is probably no need for a lifting roof except when it is necessary to trap the parents for ringing, weighing and general inspection (when the entrance hole will need to be plugged). If this method is chosen, then the roof may be attached to a suitably upward extended back wall with a piece of rubber inner tube or leather. Another advantage of the RSPB box is that by removing the entrance-hole section the box is converted to a semi-open plan layout suitable for Robins, Spotted Flycatchers and Pied Wagtails.

For hole nesters like tits, though, the entrance must definitely not be at floor level. The birds need to import a quantity of nest material which will occupy a few inches at the bottom of the box, and in any case the chicks would be extra vulnerable to predators if there was an opening at their nest cup level. There is no need for a back door or tradesman's entrance to the box. The single entrance hole must be near the top, on any side, at least 5in (125mm) from the floor, thus containing the nestlings and discouraging cats' paws which might fish them out. Tits prefer a very small entrance hole, another feature which makes theft difficult for cats. A 1⅛in (29mm) diameter hole will allow easy access for both Blue and Great Tits. Blue Tits are just able to manage a 1in (25mm) entrance, but the tolerance takes some achieving if you really want to exclude the dominant Great Tit in favour of the smaller bird. A 1¼in (32mm) hole fairly precisely defeats House Sparrows, but not Tree Sparrows, which can be equally persistent in their efforts. They may, on occasion, fill a box which has frustrated their entry with nest material, thus effectively spoiling it for the use of other birds. As the indefatigable Baron von Berlepsch said, 'where success with nesting boxes is aimed at, the fight against sparrows must not be overlooked.' A 1½in (38mm) hole excludes Starlings, which are unable to squeeze through. Once you have achieved your desired Blue Tits, don't be tempted to increase the size of the hole

Blue Tits. Only five nestlings in this box, but there could be as many as fourteen! The neat cup is soon abandoned and the material trodden into the floor of the box. *OSF/Bruce Coleman*

There is no need for a perch at the entrance to a titbox, the birds fly in or out without difficulty.
*Eric & David Hosking/FLPA*

when you see them pecking at the entrance. They are not trying to make it larger in order to get in more easily – tits just tend to peck at anything. If the holes are too large, incidentally, Nuthatches will plaster them with mud to reduce them to a size which suits.

Do not include any kind of perch, intended to allow the incoming or outgoing bird a place to rest and fold its wings, before it negotiates the hole. Study of movie film has shown that they close their wings with great facility in order to pass through, and the exterior of the box is a dangerous place where they have no desire to loll about. Perches only serve to make life easier for cats and Weasels. But a clear, uncluttered flight path to the hole is important, giving room for the spread of full wingspan right up to the entrance.

Nestbox construction does not need to be to a high standard of carpentry. The completed work must be windproof and rainproof, and if this is achieved by liberal use of sealants the birds will not complain. A poor fit around floor level has the advantage of providing vital water drainage. Inevitably water of some kind finds its way into the box, and the down feathers of nestlings provide precious little protection. The chicks are susceptible to chilling and subsequent exhaustion. Be careful to achieve a rainproof roof, and make sure drips do not find their way through the entrance hole. Bore ventilation holes in the floor section, and also one or two in the top of the wall where it gets some protection from the roof. But don't make the holes big enough to encourage a queen bumble bee or tree wasps to enter and prospect the interior as a possible hive. This means the holes must be less than 3/16in (4.8mm) in diameter unless they are covered by gauze. Queen bees do have a tendency to overwinter in nestboxes, and then proceed to found a colony. But the chance is not great, and it has to be said that the bees ought to be welcome.

Use copper or galvanised nails, or brass screws, to assemble your nestboxes. Ordinary nails will soon rust and the box disintegrate. A coat of preservative (Cuprinol or creosote) will protect softwood boxes. There seems little point in going to the trouble of painting the box, just to make it more easily seen by predators or small boys! However, natural wood colours, or browns or greens, are acceptable. A ten-year colour trial carried out by the US magazine *Audubon,* found that red was the most attractive colour for boxes, closely followed by green, then blue, with yellow well down the occupancy list. For some curious reason the test did not include natural wood boxes, or black or white ones, so I don't think we can learn much from it. On the whole it seems likely that the external appearance of the box is not very important to the bird – its concern is with the indoor facilities.

Siting the box is a critical affair. The main criteria are protection from the elements and from enemies. First consider the life-style of the intended occupant when you choose a place for it. Tit boxes should be attached to walls or trees, or anything which makes some kind of substitute for a tree, which is where they will be looking for likely entrance holes. The open plan boxes for Blackbirds, Robins and Spotted Flycatchers should be placed against a wall where they are hidden in a dense jungle of Ivy or some other creeper, or in a thick hedge, or in a fruit tree where there is some cover, and so on. They are best fixed in discreet crutch sites, invisible to the outside world. The birds will find them.

Ideally the nestboxes should be fixed in position in October or November. This gives them a chance to weather into their surroundings, and their potential occupiers have plenty of time to get used to them and to explore their possibilities. It may well be that they will be used during the winter months as a roost box. However, there are plenty of records of songbirds – and Kestrels, too – taking over a nestbox in the breeding season on more or less the day it was erected.

Most garden nestboxes should be fixed at about 5ft to 6ft (1.5m to 1.8m) above the ground. Factors such as possible disturbance will obviously affect the decision, and it is worth remembering that birds will nest at heights ranging from ground level to tree top. In order to protect them from hot sun and wet Atlantic winds, the general rule is that the opening should face somewhere in the arc from north through east to south-east, but if it is well sheltered this question of orientation is probably not significant. Nestboxes do not need to be hidden away in a dense clump of trees, or in the middle of a vast woodland. It is better to site them at the edge of a copse, or ride, at the interface between lawn or grass field and the trees. This is the kind of country which gives the best feeding return. There must be an uninterrupted flight path to the entrance, and a distinct lack of places for cats to lie in wait. A convenient staging post some 6ft (1.8m) away may be an asset. This can be anything from a clothes-line to a twig, by way of a specially stuck-in post.

Firm fixing for nestboxes is not a prime requirement, after all, birds often build nests in places which sway with the wind. Even boxes which literally hang from tree branches are successful. But for all that it probably makes good sense to fix securely, and no one, least of all the occupant, wants the box to collapse because of an inadequate screw or rusty nail. And fix the box by way of a batten which will help to ensure that it does not become permanently wet and disintegrates prematurely. Many clutches are drowned in natural nest-sites every year, so take particular care to see that rainwater cannot find its way into the box, remembering that a lot of rain trickles down a wall or tree trunk, and that it tends to follow well-established channels. Make sure the box stands proud of them, or to one side. Incline the box outwards with a slight slant, so that drips from the projecting roof do not go through the entrance hole. Do not worry if the floor of the box, when set up, it is not quite horizontal. The birds will solve this problem when they import nest material.

If you want to erect the box against a particularly fine tree causing least damage, use trenails as the old-time shipwrights did when nailing planks to frames in 'wooden walls'. These conical pegs of oak or iroko will do the job perfectly well and will not damage the tree (or the saw in due course). Alternative choices are to use plastic-covered wire and 'whip' the box to the branch (but this will have to be replaced each season to avoid constricting growth), or to use copper nails which are relatively soft and kind to the forester's saw. Otherwise use 3in (70mm) galvanised nails like most of us and grit your teeth.

It is not easy to suggest the number of boxes you should erect for any given species. The quantity depends both on the availability of natural sites and on the local food potential, natural and artificial. But first consider the life-style of your intended occupant. It is a waste of time, for example, to expect two pairs of Robins to set up house in close proximity, though it may well make sense to put up more than one box in the hope of getting one occupied. Robins are fiercely territorial, Blackbirds slightly less so, but they won't object to other species living close by. So your Robin boxes can be close to tit boxes. Tits, too are aggressive, status-seeking birds, so don't put tit boxes too close to one another. Three or four boxes to the acre is probably a useful rule of thumb.

Some birds, which are highly sociable at other times, go solo when they contemplate breeding. Starlings, for example, prefer to establish family independence at breeding time. Others remain colonial, enjoying sociability and the safety of numbers. House Martins, once they have taken to a site, tend to flock and there may be dozens of mud huts around the eaves of a favourable house. Swifts and sparrows, too, subscribe to the jolly principle of the more the merrier.

Great Tits are the most enthusiastic of all nestbox occupants, and they are not fussy about the quality of the joinery! *E. Breeze-Jones/Bruce Coleman*

Songbird nestboxes may be primed with a layer of moss or a plaited ring of straw to make them even more attractive to house-hunters, but this is unlikely to be an important consideration in determining whether the box will be 'bought' or not. But Willow Tits will only ever use a box which has been jammed solid with woodshavings or polystyrene chips, so that they can excavate the cavity themselves. Woodpeckers may just possibly be impressed with the sight of a few wood chippings at the bottom of their box.

Great Tits lay a clutch of anything up to a dozen eggs; incubated for a fortnight, the chicks are airborne in about another three weeks. *M. Jones/FLPA*

As with the Robins, and indeed many other species, the male Blue Tit has the job of finding likely sites and he then takes the female to see them. She chooses, then cleans it in preparation for nest building. The cleaning may not be very thorough, and frequently they will simply build on top of an old nest. The corners will be filled and a foundation of mosses and grasses laid. And it is at this point we can enjoy the pleasure of providing nest material from a 'builder's yard', although we must under no circumstances try to join in with the job of building. Not surprisingly, birds tend to make use of the building materials which are most conveniently at hand. Nests in trees tend to be made of twigs, those on cliffs of seaweed and driftwood, on beaches of pebbles, on moorland – heather. And in a factory – bits of wire!

If you provide suitable materials, birds will enthusiastically collect them. In the breeding season, when the scrap and peanut cages are not required for food, they can be packed with straw, feathers, dog or cat combings, *short* bits of cotton, cotton-wool, sheep wool, etc. In the garden, this may reduce the amount of thieving the sparrows and Jackdaws will indulge in when they try to unravel the string from your beanpoles, or tease out threads from your clothes on the washing-line. Tree species will prefer your offerings to be hung from the branches. But put some at ground level, in a mesh bag perhaps, firmly pegged so that they can't be carried away in bulk.

If the weather is particularly dry in May, when the House Martins are busy plastering their nestcups, it may be helpful to pour a couple of buckets of water over the earth in a likely place, eg along a dusty country lane or over a well-worn bare earth patch on the lawn or in a park. They sometimes have difficulty in finding mud puddles for which to pick up their nest gobbets.

The collection of nest material offers an opportunity to the newly engaged pair of birds to reinforce their pair bond, when they offer and exchange particularly choice pieces. In the case of Rooks, for instance, you will see them, at the chosen nest site, excitedly passing sticks to and fro with much exaggerated posturing and ritual exchanges of compliments. And, for many species, when the nest is completed the exchanges will continue, the cock bird offering choice food morsels instead of twigs in the ritual courtship feeding – a procedure which provides the female with nourishment to help form the eggs and keep her alive during the period of incubation. The hens of some species obtain more than a third of their food from their mates during the early part of the feeding season.

Whatever the type of nest, and wherever it is built, whether in nestbox or natural site, its object is the same – to create a cup which will hold the eggs in warmth and yet retain a measure of security from predators and shelter from the elements. It may range from the casual few twigs of a pigeon to the most elaborate nursery of a Goldfinch. (In the case of some seabirds and raptors, there may be no nest at all, but in those cases the eggs are well protected by other factors.)

If you *are* lucky enough to be successful in enticing a breeding pair to your nestbox, keep any inspection to a minimum. Be thoughtful with your photography, and in particular be extremely sensitive with any 'gardening' you may be tempted to do with natural nests which don't quite suit your angle of view. You may, inadvertently, betray the nest to a predator. If you are recording the career of your nests for the British Trust for Ornithology's Nest Record Scheme (see page 178), keep your visits to a minimum. The well-being and safety of the birds is paramount – put their interests first. If you want to count the young, wait a few days from hatching in the case of small birds. And don't creep up on the nest with exaggerated fairy footfalls. Better to let them know you're coming. It is also important not to disturb nests at the stage when their occupants are close to leaving, since there is a danger they might 'explode' away and become exposed too soon to the attention of the world at large. However, if this disaster should occur, collect up the chicks in a handkerchief as best you may and post them back home. Then block up the entrance for ten minutes or so, until they have quietened down.

Every year the newspapers enjoy a silly bird season where nests are photographed in every conceivable odd position. Blackbirds and Pied Wagtails have nested in every known make of motor vehicle, including aircraft, and often fledged their young successfully in spite of daily trips to the office and back! Rooks have colonised the vertical ladders on the side of refinery chimneys, seeing them as perfect substitutes for trees in an increasingly treeless landscape. And Black Guillemots nested on the Yell Ferry in the Shetlands, providing the young with daily trips long before they were scheduled to go to sea!

Unhatched eggs, dead or disappearing juveniles, are a distressing possibility. Any number of causes may account for them, apart from the natural loss to predators. A dead parent, inexperienced first-time parents who may have failed in their duty, shortage of food at a critical time – all are possibilities. Double check to see whether rainwater, or cold winds, or the sun's heat, or a cat perch, were responsible. And if you are reasonably certain the fault was not yours, reflect that a natural event of this kind is the normal end for an enormous number of nestlings. It is something which is part of the expected scheme of things, not to be dwelled upon unduly.

Many clutches of eggs and broods of young chicks will be lost to squirrels, Weasels, cats, crows, and Great Spotted Woodpeckers. The loss rate is highest in the case of open nests, particularly when it is early on in the season and they are least concealed by growing leaves. But even the most sturdily built nestbox will be the subject to attack from the ground and the air. Cats will try to hook the contents out with their paws, or lie in wait, sitting on top of the box. Weasels are common garden predators – they are good climbers and able to squirm through even a 1⅛in (29mm) Blue Tit hole to fish out the eggs or chicks. Grey Squirrels simply reach in for their reward, and if the roof or lid or removable wall-piece is not secure, they will knock it off. They may even gnaw their way in by enlarging the entrance. The only remedy for this is to plump for a concrete nestbox (with the disadvantages of weight and tendency to overheat), or to protect the entrance hole with one of the metal plates which the RSPB will supply (for address see page 182). Tying a bundle of gorse branches around the tree trunk, or even around the box itself, will be a deterrent, especially to cats. However, this remedy will not affect the nest raiding propensities of the Great Spotted Woodpecker, a persistent robber. It too will dislodge a loose lid, or enlarge the entrance by chipping until it is big enough for them to get in. The

A fledgling, hardly able to fly, clumsy and appealing, may look lost but is almost certainly not. Its parents will be nearby, waiting for you to leave before responding to its calls. It is much better to leave the fledgling alone unless it is in obvious danger, when it can be moved to the nearest cover.

In their main colour patterns, most fledglings are similar to their parents, particularly the female. So, even if you cannot immediately identify it by seeing it with its parents, it will probably be obvious whether your fledgling is a tit, a thrush or a finch.

A baby Blue Tit has lots of yellow and blue/grey and the Great Tit has the black head and bib, typical of the adult.

A young Spotted Flycatcher is greyer, paler and more heavily spotted than a Robin.

The young Robin is not at all like its parents as it does not have an orange breast and is heavily speckled.

The Dunnock fledgling has pale edges to its feathers on back, head and breast.

Like the young House Sparrow, a fledgling Chaffinch resembles its mother. It already has the bold white wing bar and shoulder patches.

The most familiar fledgling in the garden is likely to be the House Sparrow, noisily demanding food from its parents.

The Song Thrush is like a small version of the adult. It is warmer in colour than the larger, paler Mistle Thrush, and its spots are more regularly spaced in lines curving around its breast.

The three young thrushes are superficially similar.

A young blackbird is darker than the other thrushes, but the speckles on its breast can cause confusion.

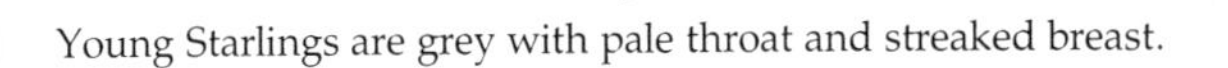

Young Starlings are grey with pale throat and streaked breast.

RSPB

metal RSPB plate may do the trick, but with woodpeckers there is the problem that they simply drill themselves a side entrance with their chisel bills to effect an entry. If they are really troublesome then a concrete box is the answer. Given the chance, these woodpeckers will take young House Martins and sparrows, as well as tits.

One other hazard which nestbox users face is that sparrows and Starlings may take boxes over from their rightful owners, or intended owners, by sheer brute force. But this is all part of the rich warp and woof of bird life.

At the end of the breeding season, in say September and October (but remember that some species, even tits very occasionally, raise more than one brood), remove the used nests and give the boxes an anti-bug spray. Do the job with caution, for used bird-nests carry risk of human diseases. The nests and box crevices will be home for feather lice, mites, ticks and flea larvae – creatures which can survive long periods without their host – so you will need to dust with a squirt of pyrethrum, or brush on an end-of-season coat of creosote or Cuprinol. Then it will be ready for a winter let. Moths may over-winter in them, perhaps even toads, mice or bats. Great and Blue Tits will certainly use them for winter warmth, roosting in solitary splendour. They often take to rather over-lit, but centrally heated, street lamps for overnight roosts. House Martins, which frequently produce three broods of young and whose breeding season will often extend into November, roost in family parties in their nest cavities, which must make for a tight fit.

Wrens are the record holders for mass nestbox occupation. Although they are reluctant to nest in the boxes, they use them enthusiastically for the winter warmth they provide. Severe winters hit them hard. Their preferred food is insects and they must work hard to get enough fuel to survive the cold nights. As small birds, with a lot of surface area relating to their volume, they suffer a great deal from heat loss and have a pressing need for warmth. At night they tend to huddle together in old nests or holes, and nestboxes suit perfectly, with the added advantage of good thermal insulation.

Thirty or forty Wrens commonly creep into their chosen roost box, and the record to date is an astonishing sixty-one, which patronised a box 4½ × 5½ × 5¾in (115mm × 140mm × 146mm) large in Norfolk. Apparently, the last arrivals were so desperate to get into the warmth that they grouped themselves together and barged in the entrance hole as a scrum – there's strength in numbers. In that astonishing case, only one of the birds was found dead in the morning, but this was presumably the effect of cramped quarters on a bird which was already in poor condition. It seems unlikely to be suffocation which causes mortality in these circumstances. The bird's metabolism slows down at night, with reduced requirement for oxygen. So even if you have a party of would-be record breakers using your tit box, don't bore any holes to increase ventilation – the Wrens are coming in because of the warmth, not the fresh air.

Give the boxes a spring clean before the breeding season, cleaning out any droppings left by the winter occupants. Another squirt of pyrethrum will kill the bugs. Prime with a small quantity of hay to offer a welcoming foundation to house-hunters at courting time.

## Birds which use nestboxes, ledges or rafts

| | |
|---|---|
| Blackbird | Moorhen |
| Blackcap | Nuthatch |
| Brambling | Osprey |
| Bullfinch | Owl, Barn |
| Bunting, Reed | Owl, Little |
| Bunting, Snow | Owl, Long-eared |
| Chaffinch | Owl, Tawny |
| Chiffchaff | Pheasant |
| Chough | Pigeon, Wood |
| Coot | Pipit, Meadow |
| Crossbill | Pipit, Rock |
| Dipper | Redpoll |
| Diver, Black-throated | Redstart |
| Diver, Red-throated | Redstart, Black |
| Dove, Collared | Redwing |
| Dove, Rock | Robin |
| Dove, Stock | Rook |
| Duck, Tufted | Skylark |
| Dunnock | Shelduck |
| Eider | Siskin |
| Fieldfare | Sparrowhawk |
| Firecrest | Sparrow, House |
| Flycatcher, Pied | Sparrow, Tree |
| Flycatcher, Spotted | Starling |
| Fulmar | Swallow |
| Gadwall | Swan, Mute |
| Goldcrest | Swift |
| Goldeneye | Tern, Common |
| Goldfinch | Thrush, Mistle |
| Goose, Canada | Thrush, Song |
| Goose, Greylag | Tit, Blue |
| Grebe, Great Crested | Tit, Coal |
| Greenfinch | Tit, Crested |
| Gull, Black-headed | Tit, Great |
| Gull, Herring | Tit, Long-tailed |
| Hawfinch | Tit, Marsh |
| Heron, Grey | Tit, Willow |
| Hoopoe | Treecreeper |
| Jackdaw | Turnstone |
| Jay | Wagtail, Pied |
| Kestrel | Waxwing |
| Kingfisher | Wheatear |
| Linnet | Woodpecker, Great Spotted |
| Magpie | Woodpecker, Green |
| Mallard | Woodpecker, Lesser Spotted |
| Mandarin | Wren |
| Martin, House | Wryneck |
| Martin, Sand | Yellowhammer |

## Exceptionally Long-lived Bird Individuals

Blackbird 14
Blackcap 7
Brambling 8
Bullfinch 9
Bunting, Reed 9
Chaffinch 11
Chiffchaff 6
Chough 16
Coot 13
Crossbill (3)
Crow, Carrion 13
Cuckoo 9
Dipper 8
Dove, Collared 16
Dove, Rock (6)
Dove, Stock 9
Duck, Tufted 15
Dunnock 9
Eider 27
Fieldfare 9
Firecrest (1)
Flycatcher, Pied 8
Flycatcher, Spotted 9
Gadwall 21
Goldcrest 4
Goldfinch 7
Goldeneye (4)
Goose, Canada 20
Goose, Greylag 18
Greenfinch 11
Gull, Black-headed 25
Gull, Herring 26
Hawfinch 7
Heron, Grey 18
Hobby (3)
Hoopoe (2)
Jackdaw 14
Jay 16
Kestrel 15
Kingfisher 4
Kittiwake 23
Linnet 8
Magpie 15
Mandarin 6
Mallard 20
Martin, House 6
Martin, Sand 7
Merlin 12
Moorhen 11
Nuthatch 8
Osprey 10
Owl, Barn 13
Owl, Little 10
Owl, Tawny 21
Pheasant (4)
Pigeon, Wood 15
Pipit, Rock 8
Redstart 8
Redstart, Black 4
Redwing 7
Robin 8
Rook 18
Shelduck 20
Siskin 6
Skylark 10
Starling 15
Sparrow, House 12
Sparrow, Tree 6
Sparrowhawk 11
Swallow 9
Swan, Mute 24
Swift 16
Tern, Common 24
Thrush, Mistle 11
Thrush, Song 10
Tit, Blue 12
Tit, Coal 8
Tit, Great 13
Tit, Long-tailed 8
Tit, Marsh 9
Tit, Willow 10
Treecreeper 8
Turnstone 17
Wagtail, Grey 6
Wagtail, Pied 9
Warbler, Willow 9
Warbler, Garden 7
Waxwing (2)
Wheatear 7
Woodpecker, Great Spotted 10
Woodpecker, Green 15
Wren 5
Wryneck (2)
Yellowhammer 11

**Further reading:**
There is more information about the science of nestbox materials, construction, fixing, siting and preserving in the definitive booklet *Nestboxes* by Chris du Feu, published by the BTO (for address see page 182).

## LIFESPANS OF WILD BIRDS

These figures are for the oldest ringed wild individual marked in Britain and Ireland by the British Trust for Ornithology. Over twenty million birds have been ringed to find out about their movements and survival. If you find a ringed bird – dead in the road or brought in by the cat for instance – write to the BTO (see address page 182) with the details. These should include the ring number (and address if not British), where and when you found it, what had happened to the bird, what species you think it was, and if dead whether it was fresh, old or mummified. Finders are always told about the origins of the bird they report, so include your own address. The only exception is in the case of racing pigeons (their rings are usually encased in plastic and not split), when you should write to the Royal Racing Pigeon Association (address page 183).

The lifespans listed opposite may be for the very oldest bird from tens of thousands reported (like the Blue Tit at 12 years) or from relatively few records of a species rarely ringed (such as the two year old Hoopoe). If the figure is surely unrealistic as so few have been ringed it is in brackets. The figure is the age between ringing and finding for birds that were either alive or freshly dead when they were discovered. The oldest bird from BTO ringing was a Manx Shearwater, a globe-trotting seabird which spends the winter off South America. This was ringed on 31 August 1954 and was last re-trapped by its ringers, at its colony off County Down, in August 1987. It's probably still alive!

The British Trust for Ornithology runs a Nest Record Scheme, in which the object is to collect information about the breeding behaviour of British birds. If you feel you could collect simple, but precise, information about the birds which nest in your garden, and not necessarily in nestboxes, write to the BTO for information (for address see page 182).

# SPECIES NOTES

These notes provide basic information about those birds which patronise artificially-provided nest places. Status and distribution, habitat, food and feeding, nest and nesting is summarised. There is no information on identification, since this would run away with too much space and I am assuming that a field guide is essential to any bird gardener's library.

One of the best reference books for bird identification in the British Isles is *The Hamlyn Guide to the Birds of Britain & Europe*, by Bertel Bruun. For identification notes plus detailed information on habitat, distribution, behaviour, food and breeding (in the wild) the most useful single volume is *The Popular Handbook of British Birds*, by P. A. D. Hollom, published by H. F. and G. Witherby. Further information on nestboxes can be found in the pamphlet *Nestboxes*, published by the BTO (for address see page 182).

This list is assembled in the internationally accepted Voous order of classification, where the birds are arranged in a sequence of related families which makes taxonomic sense. An alphabetical list is provided in the general index at the back of this book.

NB 'Hole height' is the distance from the floor of the box to the *bottom* of the entrance hole. 'Floor' is the enclosed area *inside* the bottom of the box.

Additions and corrections to the food preference and nestbox sections in the following notes will be welcomed.

**Red-throated Diver** *Gavia stellata*

Local in Scottish Highlands, islands, Orkney, Shetland and Donegal. Pools, lochs, marsh and moor. Outside breeding season is marine.

Fish, maybe molluscs and crustacea, etc.

Nests close to water, usually on islets, sometimes at lochside. A mere scrape in the vegetation; sometimes a platform of vegetation.

Artificial nest site: In Argyll, where the divers were suffering from a good deal of disturbance both by fishermen and hydro-electric scheme water fluctuations, local enthusiasts found that the birds would accept a raft as a nest site. Empty plastic containers topped with heavy-gauge wire netting, planted with turves and bound with yet more wire netting, were anchored in suitable hill lochs. Avoid sites where the birds already breed successfully, but choose places where they have bred in the past or show themselves freely.

Eggs: Usually 2 yellowish-olive to brown. Late May or June. Incubation 24–29 days, fledging about 8 weeks. One brood.

**Black-throated Diver** *Gavia arctica*
Local in Scottish Highlands. Large lochs.
*See* Red-throated Diver.

**Great Crested Grebe** *Podiceps cristatus*
Breeds regularly in most English counties except Devon and Cornwall. Scarcer in Scotland and Wales. Lakes, reservoirs, gravel pits and large ponds with reedy cover.
Dives for fish, insects, tadpoles, etc.
Naturally nests among reeds or vegetation close to edge, made of water plants, reeds, perhaps twigs, just above surface.
Artificial nest site: May adopt the sort of raft put up by wildfowl enthusiasts for geese and ducks (see page 81). Needs a reasonably stable base on which to build a heap of vegetation.
Eggs: Usually 3–4 chalky white, become grained during incubation. End March onwards. Incubation 28 days; fledging 9–10 weeks. Sometimes two broods.

**Fulmar** *Fulmarus glacialis*
Summer visitor, breeding on coastal cliffs more or less round the whole of the British Isles. Otherwise at sea in the North Atlantic.
Feeds on molluscs, fish, etc. Takes fish-offal and 'trash' thrown overboard from fishing vessels.
Nests on cliff slopes and ledges, on bare rock or soil. Sometimes the female makes a slight hollow.
Artificial nest site: Has taken to excavated ledges provided in Norfolk cliffs, along stretches where there are few natural sites. On exposed cliffs, where there are patches of sand in the boulder clay, dig out a ledge about 1ft (300mm) wide and 6in–8in (150mm–200mm) deep in the sand. The birds do the rest.
Eggs: White. Late May. Incubation 8 weeks; fledging 8 weeks. One brood.
Read: *The Fulmar* by James Fisher, Collins, 1952.

**Grey Heron** *Ardea cinerea*
Resident throughout British Isles, wherever there is water not too deep for wading.
Mainly fish, but much else. Fish farmers objecting to heron visitors should contact the RSPB. May come to bird table for kitchen scraps, will come to garden pond for Goldfish, etc.
Nests in tree canopies, colonially. Single nests sometimes found which may signal the founding of a new colony. Bulky structure of branches, sticks, lined with smaller twigs.
Artificial nest site: May take advantage of a platform on chicken wire frame firmly placed high in Scots Pine or other suitable tree.
Eggs: 3–5 greenish-blue. February or March. Incubation about 25 days: fledging about 50–55 days. Sometimes two broods.
Read: *The Heron* by Frank A. Lowe, Collins, 1954.

**Mute Swan** *Cygnus olor*
Generally distributed. Open water, ponds, parks, sheltered estuaries, sea coast and lochs.
Dips head and neck, or 'up-ends' to graze on underwater vegetation; also takes roots and buds of aquatic plants, small frogs, tadpoles, fish. Will come to hand or feeding station for bread.
Nests anywhere near water, on large heap of vegetation.
Artificial nest site: Takes readily to a suitable raft, both on freshwater and on estuaries provided there is a gentle slope with easy access. Prime with a pile of vegetation.
Eggs: 5–7 almost white, tinged with greyish- or bluish-green. April or May. Incubation about 35 days; fledging about 4½ months. One brood. Warning: aggressive at nest.
Read: *The Royal Birds*, Lillian Grace Paca, St Martin's Press, New York 1963; *The Swans*, Sir Peter Scott and The Wildfowl Trust, 1972; *The Mute Swan*, John Fair, Croom Helm, 1986.

**Greylag Goose** *Anser anser*
In summer, hilly heather moors, islands. Feral birds breed more freely by freshwater sites such as reservoirs, gravel pits and lakes, mainly on islands.
Food: Grasses, cereals.
Nests on the ground. Heather or twigs, grasses, mosses, with down and feathers.
Artificial nest site: Has regularly nested on rafts in suitable locations.
Eggs: 4–6 creamy white. Last half of April. Incubation 27–28 days; fledging about 8 weeks. One brood.
Read: *Wild Geese,* M. A. Ogilvie, Poyser, 1978.

**Canada Goose** *Branta canadensis*
Was introduced to Britain as a status symbol, ornamenting stately lakes, in eighteenth century. Has since become a successful feral species, enjoying grassland and marshes by freshwater ponds and lakes.
Grazes in flocks on grassland. Also takes water plants. Will come to hand-feed on corn or bread when tame.
Nests on islands and marshes, sheltered by undergrowth or bush. Nest-hollow lined with grasses, leaves, reeds, down and feathers.
Artificial nest site: Box or platform raised on posts above water level or on raft. Make an artificial island, plant clumps of iris, reeds, sedge, etc to provide a nest site. Since the expansion of gravel pit workings of the last few decades they have taken advantage of the soil islands which remain when the pits are worked out and flooded. In Canada, where they sometimes nest in suitable tree sites such as broken stumps or in the hollows left by fallen branches, this propensity has been exploited by egg collectors. Wooden platforms, up to 65ft (20m) above ground, or on top of 10ft (3m) poles where there are no trees, are soon colonised. Try sawn-off barrels or open tubs, suitably drained, and offering a platform some 2ft (600mm) across and 1ft (300mm) deep.
Eggs: 5–6 white. Late March or April. Incubation 4 weeks; fledging 6 weeks. One

The provision of nestboxes has encouraged a remarkable increase in the British population of Goosanders. *Hannu Hautala/FLPA*

A 'rabbit burrow' nest site attracts Shelducks.

brood. Warning: Canada Geese (gander especially) can be very aggressive in the breeding season, and have been known to attack human beings, even wounding children.
Read: *The Canada Goose*, Kit Breen, Swan Hill Press, 1990.

**Red-breasted Merganser** *Mergus serrator*
May take to a duckbox, 24in × 18in × 18in (600mm × 450mm × 450mm), entrance hole 8in (200mm) square.

**Goosander** *Mergus merganser*
Large box, with an entrance hole 4in (100mm) in diameter placed fairly high and with a clear line of flight to the water.

**Shelduck** *Tadorna tadorna*
Generally distributed round low-lying coast and estuaries.
Food: Marine molluscs, crustaceans, etc.
Nests in Rabbit burrows, bramble tunnels, in gorse and Bracken, sometimes in walls, hollow trees. Lined with down plus some vegetation.
Nestbox: Ideally on an island free from predators. Plastic pipe or old milk churn set in bank. May be persuaded to nest in a fruit box in a hollow in vegetation, for instance Wild Rose or bramble bushes surrounded by long grass or Bracken. Or sink a small barrel into the ground with just a 6in (150mm) entrance hole showing. Provide a 1ft (300mm) cube nest chamber approached by a length of 9in (230mm) drain pipe. Simulate a Rabbit burrow. Alternatively, make a wigwam of twigs in undergrowth.
Eggs: 8–15 creamy-white. May. Incubation 28 days; fledging 45 days. One brood.
Read: *Ducks of Britain and Europe*, M. A. Ogilvie, Poyser, 1975.

## A SHELDUCK NEST CHAMBER

A Shelduck nest tunnel.

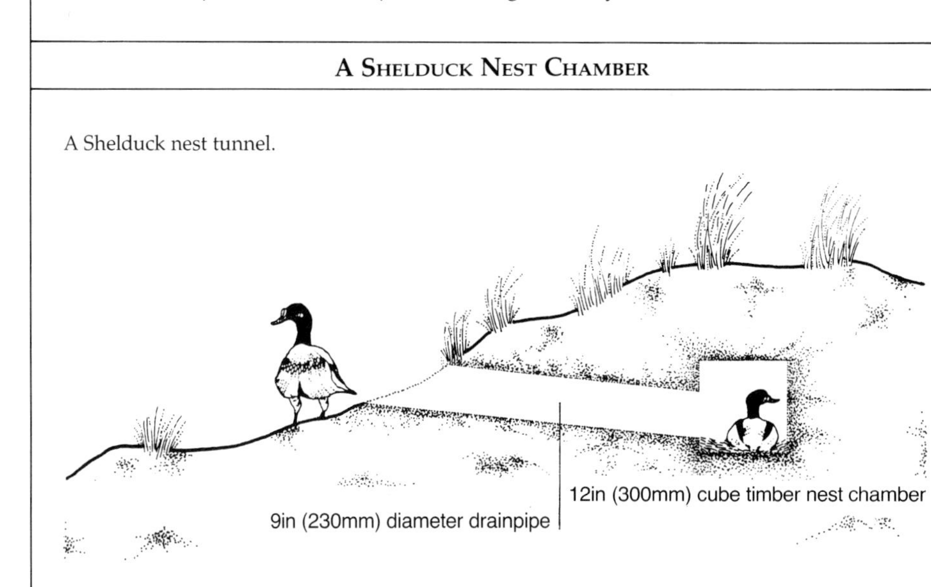

**Mandarin** *Aix galericulata*
An exotic dabbling duck which has escaped from waterfowl collections to flourish mainly in areas of Surrey and Berkshire, but increasingly elsewhere.
Surface and up-ending feeders on vegetation, but mainly graze ashore. Will come to corn on lawn. If Wood Pigeons make their lives impossible, try offering it in the pond.
Nest in tree holes, mostly oak and ash, usually 5ft–24ft (1.5m–7.2m) above ground. Reluctant to take to boxes at first, needs secluded pond or stream with trees nearby.
Nestbox: Upright box, 20in (500mm) high, 8in–10in (200mm–250mm) wide and deep with a 4in (100mm) entrance hole (any shape). Line with shavings or rotting wood loose enough for bird to hide eggs. Fix at 15ft (4.5m) height, but can be lowered in subsequent seasons if successful.
Eggs: 9–12 eggs in March, hatch in May, one brood.

**Gadwall** *Anas strepera*
Breeding bird in Britain since mid-nineteenth century, having been introduced to East Anglia. Scattered over British Isles, breeds by shallow, lowland sheets of freshwater, lakes, meres, reservoirs, marshes and slow-flowing streams. Spreading slowly.
Food: Mostly vegetable, some animal.
Nest site is concealed in dense vegetation close to water, tussocky grass, sedge, nettles.
Artificial nest site: May nest on rafts, if provided with good growth of vegetation.
Eggs: 8–12 creamy-buff. May or early June. Incubation 27–28 days; fledging 7 weeks. One brood.

**Mallard** *Anas platyrhynchos*
Generally distributed, near all kinds of freshwater, estuaries and coastal islands.
Food: Mainly vegetable. Enjoys soft potatoes.
Nests in thick undergrowth sometimes far from water. Pollard willows, tree holes, second-hand crow nests, etc. Grass, leaves, rushes, feathers, down.
Nestbox: Try providing an apple-box or large, open cat basket in typical nesting area. Where Mallards have become very tame (village ponds and the like), try erecting an open-ended barrel on an island. Otherwise a mere hollow in the ground, bordered by a couple of short logs and sheltering under a wigwam of spruce boughs, may do the trick. Mallard nests are probably best sited on rafts or islands, where they enjoy some protection from Foxes and rats.

Alternative nestbox: Using sawmill offcuts, make a box with inside dimensions of 1ft (300mm) square and 9in (230mm) high. Prime with an inch or two (25mm–50mm) of woodshavings. Make a tunnel about 1ft (300mm) long leading to an entrance hole 6in (150mm) square. This tunnel entrance serves to deter crows. A ramp should lead gently down from the tunnel entrance to the ground. This ease of access is important, not only for the comfort of the duck, but because she might take broods back to the safety of the box at night for the first couple of weeks after

leaving the nest, especially in cold weather. The duck likes to be able to see out from the nestbox so provide a horizontal slit in the side.
Eggs: About 12 greyish-green or greenish-buff, occasionally a clear pale blue. February onwards. Incubation 4 weeks; fledging 7½ weeks. One or two broods.
NB A duck-box may well be taken over by Moorhens.

**Tufted Duck** *Aythya fuligula*
Local but fairly widely distributed, except in the south-west.
Lakes, lochs and reservoirs.
Feeds mostly on insects, molluscs, frogspawn, tadpoles and frogs, and some aquatic vegetation.
Nests close to water in tussocks of sedge or rushes.
Artificial nest-site: May occasionally nest on rafts.
Eggs: 6–14 greenish-grey. Second half of May, June. Incubation 23–26 days; fledging 6–7 weeks. One brood.

**Eider** *Somateria mollissima*
Resident, breeding round Scottish coast, the north-east and north-west (non-breeders further south in summer). Rocky and sandy coasts, sea-lochs and estuaries, hugs coast, seldom inland.
Feed in company, shallow dives for seaweed, crabs, shells.
Like to nest *by* something, so offer them a stick or a fish-box.
Nest lining of eiderdown.
Eggs: 4–6 large cream eggs. End May, beginning June. Incubation 4 weeks. Young join crêches and fledge at 2 months.

**Goldeneye** *Bucephala clangula*
Established in Scotland and increasing with many of those present using nestboxes.
Food: Small invertebrates.
In Scandinavia, Goldeneyes nest in tree holes and stumps, beside lakes and ponds in thickly wooded country. In the late Middle Ages they were farmed for eggs by Lapps who improved natural nesting-sites.
Nestboxes: Like most hole-nesters, from Blue Tits to Tawny Owls, Goldeneyes take readily to artificial nests. They returned to breed in Scotland in 1971, although boxes had first been erected twenty years before. Scottish highland lochs offer ideal habitat, but nest sites are scarce in the wild, because so much of today's forestry is too tidy, depriving trees of the chance to decay and offer holes. The RSPB has put up about four hundred boxes and others have been erected by birdwatchers and foresters. The more the merrier. Materials are as follows: Rough sawn timber 9in wide and ½in thick (230mm x 12mm) cut in 9ft or 6ft (2.75m or 1.83m) lengths. Cut the lengths from the saw mill into three different sizes – work out the correct ratio for the number of boxes to be constructed. The lengths are 18in (460mm) for the sides, 8in (200mm) for the bottoms and 12in (305mm) for the lids (see diagrams). Also required are lengths of 1¼in × 1¼in (32mm × 32mm) rough wood cut into 8in (200mm) lengths for strengthening inside the boxes. The

box is nailed together. Cut 4½in (115mm) diameter entrance holes. Finally, cover the lid with 12in (305mm) wide bitumastic flashing (as used for finishing off a tarred roof) which you can buy in rolls from builders' merchants. The flashing is burnt onto the box with a blow-lamp and the sides are folded down to cover all the joints. The boxes can be treated with Cuprinol or creosote (if the latter, they need to be treated months before use). It is important that the inside of the box is rough wood and not smooth or else the ducklings cannot get out easily. There is no need for a landing perch; like tits, the ducks fly straight in.

## GOLDENEYE NESTBOX

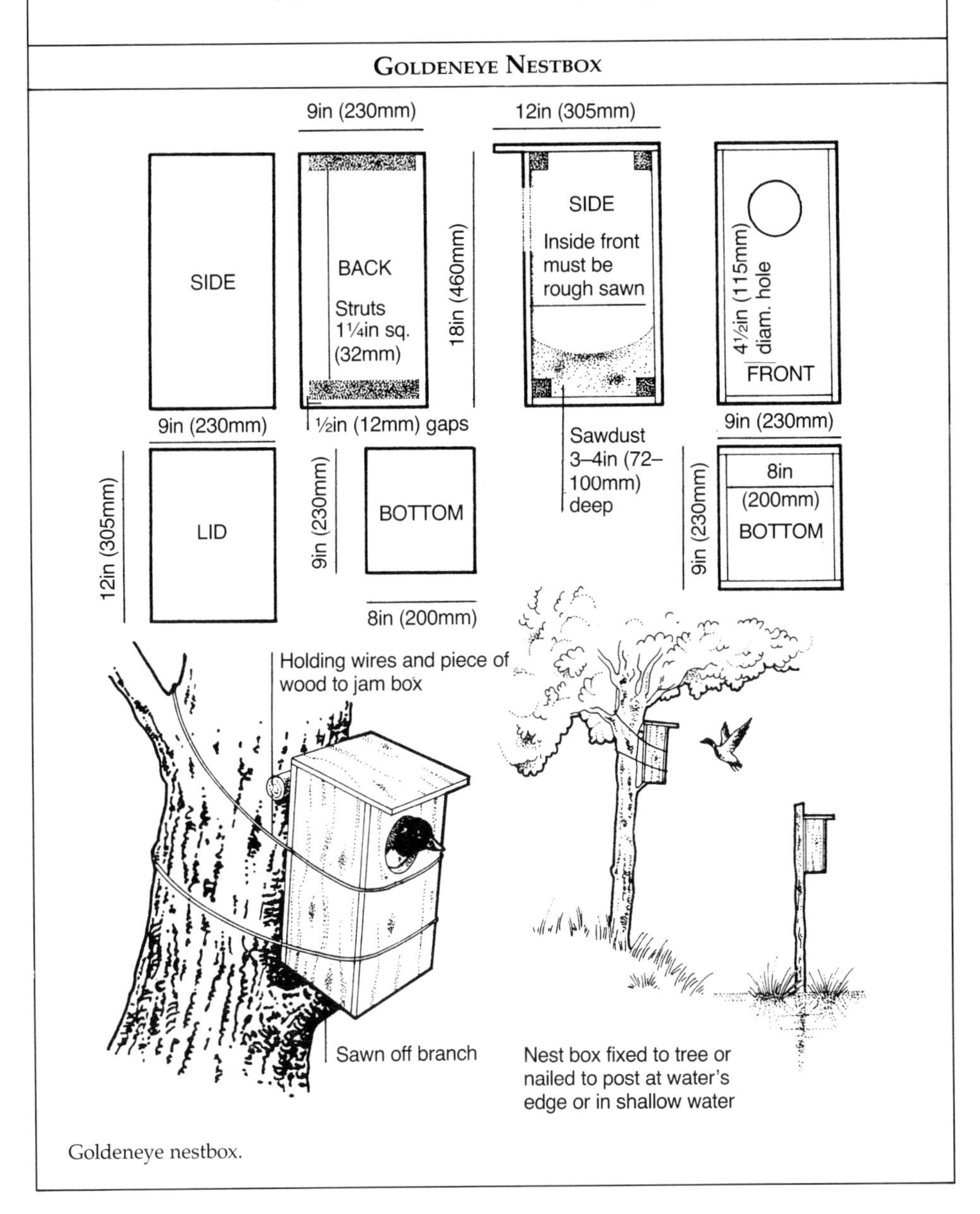


Goldeneye nestbox.

Initially boxes should be erected on trees facing open water at a height of about 10ft–20ft (30m–61m) from the ground although they can be used as low as 3ft (9.2m). The RSPB has tended to use places which are not in direct sunlight in the middle of the day and some of the boxes can be up to 30yd (27.5m) from the edge of the water so long as there is an uninterrupted view of the water and a direct flight path. Once birds have started to nest, new boxes can in fact be up to ½ mile (75m) from the water as the ducks are very good at finding nesting holes and also at leading their young to the water.
Boxes should be well spaced out and erected in small groups of 3–4; there should not be too many boxes in one area as this leads to increased competition for nest sites and may result in desertion. Quiet undisturbed bays are best but it is also worth using river banks and islands. Use a ladder to put up boxes in trees which are difficult to climb, to save them being looked into by everyone who passes by. Choose quiet sites. In Pine Marten localities, it is best to use trees on islands or growing in wet areas.
Attach boxes firmly to the trees. One of the best materials is old telephone cable. Wind it around the top and bottom of the boxes and round the tree and then tension by jamming a piece of branch behind the box. This also allows you to slant the box at an angle so that the rain can run off the lid quickly. It may be useful to cut off a branch about 6in (152mm) from the trunk to rest the box on before tying it to the tree.
Put about 3in–4in (72mm–100mm) of sawdust into the bottom of the boxes – the birds will not use the boxes without adequate sawdust. Goldeneye ducks usually lay their eggs in April/early May and the first young usually appear on the water in May/early June. The birds should not be disturbed while they are incubating (the species is on Schedule 1 of the Bird Protection Act). A check can be made to see if any of the boxes have been used by Goldeneye (there will be down and egg-shells in the nest). All boxes should be checked in late winter so that they can be fastened if they have become loose. Sometimes boxes are used by Jackdaws which carry sticks into them and these need to be cleaned out annually. Perseverance is required as Goldeneyes sometimes take a long time to move to new areas.
Eggs: 6–15 bluish-green. Mid April on. Incubation 26–30 days; fledging 57–60 days. One brood.
NB The Goldeneye boxes have been developed by Roy Dennis (Highlands Officer of the RSPB) and he welcomes information on their success (Munlochy, Ross & Cromarty, IV8 8ND).

**Goshawk** *Accipiter gentilis*
Prefers to nest in mature forest. May take to a disused crow or pigeon nest. Artificial nest platforms have been developed by the Forestry Commission.
Read: *Goshawks, their Status, Requirements and Management*, S. J. Petty, HMSO, 1989.

**Sparrowhawk** *Accipiter minus*
May visit bird table to sample small birds. Be philosophical! Sparrowhawks have to make a living!

**Osprey** *Pandion haliaetus*
Takes freely to cartwheels or ledges on top of tall poles in USA. Roy Dennis, RSPB's Highlands Officer, has had a good deal of success in 'gardening' conifers to provide an open platform some 6ft (2m) in diameter. Main problem is vandalism and egg-thieves, so try it only in well-protected areas.
Read: *The Osprey*, Alan Poole, Cambridge University Press, 1989. *Ospreys*, Roy Dennis, Baxter, 1991.

**Kestrel** *Falco tinnunculus*
Resident, generally distributed, except in winter in far north.
Moors, coast, farmland and open woodland, suburbs and cities.
Perches on trees, posts, wires or buildings, watching out for its prey. Hunts in the

A PROVEN DESIGN OF KESTREL BOX

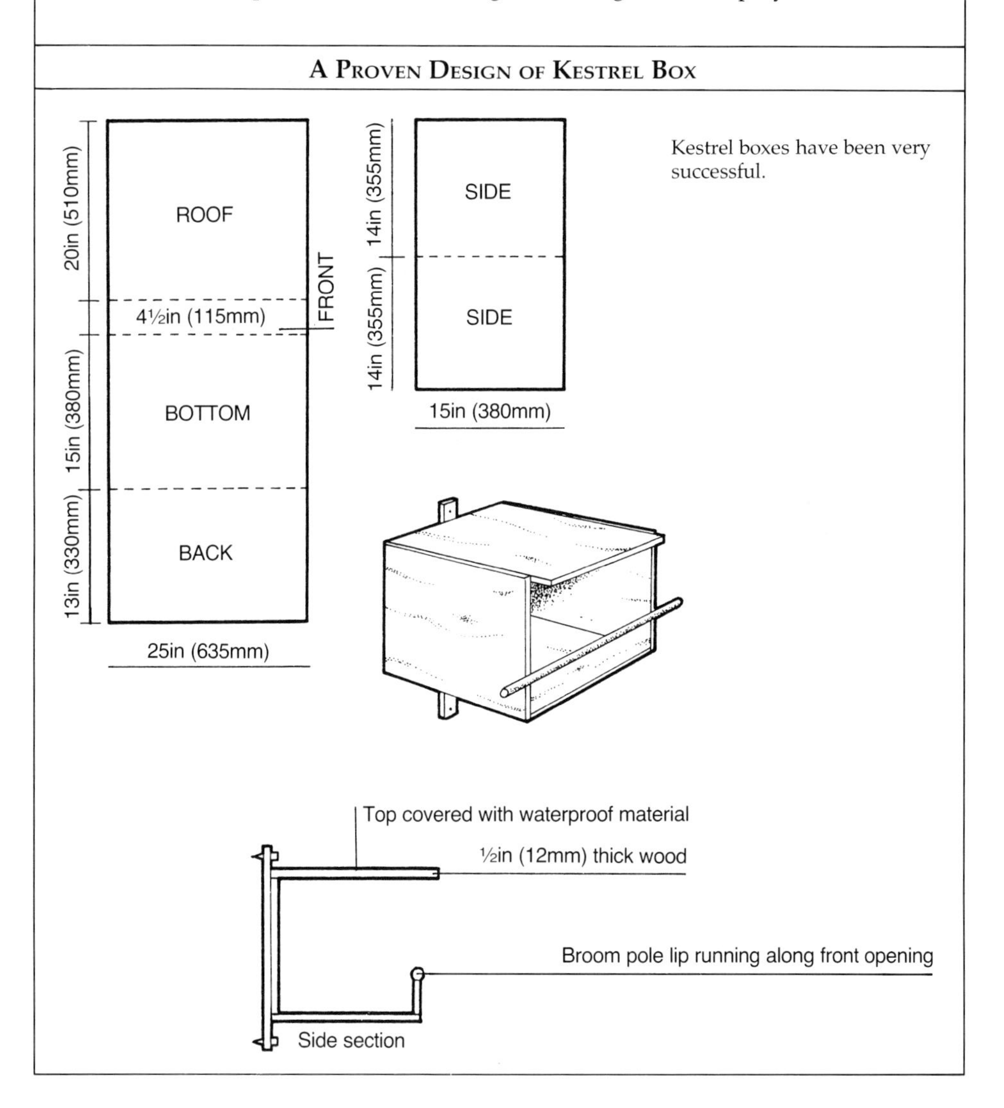


Kestrel boxes have been very successful.

Artificial nest sites are readily accepted by Ospreys in the United States and they are already proving successful in Britain. *Tony Soper*

open, checking frequently to hover in characteristic attitude, watching for beetles or small mammals. Untypically, has been known to come to bird tables to pick at turkey carcasses and day-old chicks, even broken dog biscuit.
Makes no nest, but uses a scrape on cliff or quarry ledge or uses secondhand crow nest as platform. Sometimes in tree hollow or ledge on building or ruin.
Nestbox: Open-fronted, 25in × 15in × 15in (635mm × 380mm × 380mm) high, with roof overhanging a couple of inches (50mm). One of the long sides partly open, having only a 5in (125mm) board along the bottom part, fitted with a broom pole lip to enable the bird to perch easily before entering. Prime the box with a little peat mould or woodshavings. Fix very firmly on 18ft–30ft (6m–10m approx) pole, or high on side of house where some shelter is available from midday sun. If fixed to tree, make sure chick thieves cannot climb to it easily, and place it so that wing-exercising juveniles can step out onto a branch (or extend the broom pole). Swiss-erected boxes were positioned near farm buildings, and in one year no fewer than 26 out of 36 were occupied in an area of about 5,000 acres. With the continuing loss of hedgerow hollow-tree sites it is much to be hoped that these boxes will become more popular with farmers. But it makes a conspicuous nest site, and unfortunately young Kestrels have a ready, though illegal market, so be very careful that they are well protected. Farm buildings, private parkland, nature reserves and sewage farms are ideal sites. In Holland, where farmers encourage Kestrels in controlling voles and shrews, the boxes have been highly successful.
Little Owl, Jackdaw, Collared Dove, Stock Dove or Blackbird may also use them.
Eggs: About 5, the white colour often hidden by red-brown splotchings. Mid April onwards. Incubation 28 days; fledging 28 days. One brood.

**Merlin** *Falco columbarius*
Partial migrant. Open moorland with some trees, coastal dunes and cliffs.
Feeds on small birds, some insects and small mammals. Might be seen chasing small birds near bird table if you're lucky.
Nests on ground among heather or on sand or in old crows' nests. Will take to introduced crow nests in trees at edge of open heather moorland.
Platform nest: Saucer-shaped cup of chicken wire about 20in (500mm) in diameter, interleaved with willow to strengthen the cup, then lined with an inverted sod. Site one example on ground amongst heather at edge of woodland where crows are established, site another a few yards inside the wood, as if it were a crow's nest in a tree.
Eggs: 2–3 yellowish-brown, speckled reddish-brown. Late May to late June. Incubation 28 days; fledging 4 weeks.

**Hobby** *Falco subbuteo*
Has been persuaded to take to flat baskets placed near the tops of tall trees (in Germany) and to old crows' nests suitably sited.

**Pheasant** *Phasianus colchicus*
Resident and generally distributed except in Ireland. Much-preserved as a game bird. Woodland edge, cultivated land, parkland, large secluded gardens and shrubberies, damp, rushy and sedgy fields.
Forages on ground for varied selection of animal and vegetable food. Fruit, seeds, grain, insects, worms, slugs. Will come to secluded garden ground station for corn.
Nests under cover of ferns, Brambles, etc, in woods, copses, hedgerows and reed-beds, making a hollow in the ground and lining it with a few stems of grass and dead leaves.
Eggs: 8–15 olive-brown eggs. Early April onwards. Incubation 22–27 days; fledging 12–14 days. One brood.

**Peregrine** *Falco peregrinus*
Nests on the ledges of steep cliffs, conveniently close to its preferred prey, pigeons. Logically taking to windowsills or high-rise church ledges in cities where, again, pigeons are plentiful.

**Moorhen** *Gallinula chloropus*
Generally distributed, scarcer in northern Scotland. Almost any freshwater from a ditch to a lake; fresh-water ponds, slow streams, marshes and water-meadows.
Forages on grassland and waterside vegetation. Food is mostly vegetable, but includes a fair proportion of animals, such as worms, slugs, and snails. Will soon get used to coming to scraps on the ground; flies less readily up on to the bird table. Watch out for rats.
Nests typically in shallow, still water. Platform of dead plants amongst aquatic vegetation, in trees and bushes.
Nestbox: Will happily take over a duck box, or a mini-island.

Eggs: 5–11 whitish-grey to buff or greenish eggs. April onwards. Incubation 19–22 days; fledging 6–7 weeks.
Usually two broods, frequently three.
Read: *A Waterhen's Worlds*, H. Eliot Howard, Cambridge University Press, 1940.

**Coot** *Fulica atra*
Generally distributed, except in highlands. Lakes, large ponds, slow-flowing rivers and backwaters.
Food: Aquatic vegetation, also grazes on land.
Nest is large platform of vegetation built up well above water level, among reeds or in open.
Artificial nest site: Has regularly nested on rafts.
Eggs: 6–9 stone-coloured, spotted dark brown. Second week March onwards. Incubation 21–24 days; fledging about 8 weeks. One, sometimes two, occasionally three broods.

**Turnstone** *Arenaria interpres*
Winter visitor to rocky or pebbly coasts, although many non-breeders stay for the summer.
Roots about in small parties over rocks or shore, searching for small seashore animals in weed and tide-line debris. Comes eagerly to regular beach feeding station for bread, cake, peanuts and scraps.
Breeds from Arctic Circle to points north.

**Black-headed Gull** *Larus ridibundus*
Resident and widely distributed. Coastal and inland species frequenting lakes, sewage-farms, harbours and farmland.
Feeds around low-lying shores and estuaries, freely inland to farms, lakes and rivers. Food very varied, animal and vegetable. Will come fairly freely to bird feeding stations in open situations.
Nests in colonies among sandhills, sandbanks, lake islands and shingle. Rough nest of vegetable matter. Takes freely to artificial islands.
Eggs: About 3 eggs, light buffish-stone to deep umber brown, splotched dark blackish-brown. Mid-April onwards. Incubation 22–24 days; fledging 6 weeks. One brood.

**Herring Gull** *Larus argentatus*
Resident, generally distributed along coasts, estuaries, waters and fields often far inland.
Opportunist feeder, eating almost anything, but mostly animal food. Will come to bird table or ground station for almost anything, but is shy and not particularly welcomed by other birds.
Nests in colonies on cliff ledges, grassy coastal slopes, sand dunes and shingle. Large nest of grass or sea-weed. Of recent years it has taken to nesting on roofs and chimney pot areas where scraps are freely available. Not to be encouraged, though, as it can be aggressive in defence of its young.

Eggs: About 3 eggs, olive to umber, sometimes pale blue or green, splotched with deep blackish-brown. End April to early June. Incubation 26 days; fledging 6 weeks. One brood.
Read: *The Herring Gull's World*, Niko Tinbergen, Collins, 1953.

**Kittiwake** *Rissa tridactyla*
Nests in colonies on precipitous sea-cliff ledges. Enjoys man's hospitality by colonising cliff-ledge-like warehouse window-ledges. Also inside motor-car tyres hung as dockside fenders. Needs shoe-box size ledge on vertical surface overhanging water.

**Common Tern** *Sterna hirundo*
Breeds colonially round most of British coast and inland in Scotland and Western Ireland.
Food: Mainly Sand Eels. Has been known to take bread in company with gulls feeding in the wake of a ferry (Torpoint, Devonport).
Nests in low-lying sandbanks or shingle beaches, low rocky islets and skerries. Inland on islets in lochs and low moorland, river shingle banks.
Artificial nest site: Islands or rafts, clear of vegetation. The Merseyside Ringing Group moored a raft on a reservoir at Shotton belonging to the British Steel Corporation (for details see page 81). While the surroundings might have been incongruous, the raft was effective in providing a nesting place and an old-established ternery was able to maintain its presence, thus reversing a trend towards decline (2,500 terns were reared there by 1986). To be worth all this effort, though, the raft must be close to a food source for the terns.
Eggs: Usually 3 stone-coloured. Late May or early June, in a hollow. Incubation 21–28 days; fledging 4 weeks. One brood.

**Little Tern** *Sterna albifrons*
Highly vulnerable to human trampling and Foxes, but also to flooding by high tides. Wardens at Gibraltar Point Bird Observatory, on the Wash, have developed a method for protecting Little Tern nest scrapes from predatory Foxes, while also solving the flooding problem. It involves gradually lifting the nest-shingle, eggs and all, till it is safe in a fish box, supported five feet off the ground by fence posts. After hatching, the shingle is replaced by tide-line debris which provides some protection from the elements.

**Black Tern** *Chlidonias niger*
In Holland they take freely to small rafts, provided they are in sheltered waterways, preferably surrounded by aquatic vegetation.

**Rock Dove** (feral pigeon) *Columba livia*
Resident, but decreasing in numbers and hopelessly interbred with its own descendants, domestic and racing pigeons. The only pure Rock Doves that remain are probably to be found on the north and west coasts of Scotland and Ireland. Found by rocky sea cliffs and coastal fields, foraging on coastal

Herring Gulls find rooftops an acceptable substitute for cliff slopes, especially when there is a fishing port close by. *Tony Soper*

Above: Like many birds which normally nest in tree holes, Stock Doves will take over a box, but they need a somewhat bigger entrance hole than tits. *Mike Read/Swift Picture Library*

*Below left:* Little Terns are scarce breeding birds in Britain, they need all the help they can get. This one is enjoying the view at Gibraltar Point Field Station, Skegness, Lincolnshire. *R. B. Wilkinson*

pastureland for grain, peas, beans, potatoes, seeds, etc. (The feral pigeon breeds freely in cities and places where it can take advantage of man's soft heart.)
Nests in sea-caves or among rocks at wilder parts of coast. Few bits of heather or roots in a hole or cave ledge or crevice.
Nestbox: This species was domesticated hundreds of years ago in Scotland. Artificial nesting ledges were provided for them in convenient sea-caves, and the resulting squabs farmed for food. The flourishing domestic form (feral pigeon) has confused the wild status of the bird in no uncertain terms. But whether you are dealing with street pigeons or fantails the principles of nestboxing are the same. Pigeons are happiest in a dark chamber which recalls the cave crevices of their ancestors. A garden dovecot, round or octagonal, should be mounted on a stout pole to discourage cats and rats. A two-storey structure makes sense, allowing a number of pairs to breed in companionable proximity. The 'pigeon holes' should provide chambers roughly 24in × 18in × 18in (600mm × 450mm × 450mm) high, with an entrance hole 6in × 6in (150mm × 150mm). The house should be draught-free, but well ventilated, and there should be a generous landing shelf outside the entrance holes (unlike Blue Tits, pigeons *do* like to land outside the entrance). A roof should keep rain off and also slope south, providing a warm place for the birds to sunbathe and posture.
Eggs: 2 white. April onwards. Incubation 17–19 days: fledging 4–5 weeks. Two or three broods, maybe more.

**Stock Dove** *Columba oenas*
Resident and well distributed, except in northern Scotland. Open parkland, wooded country, cliffs and sand dunes.
Feeds over fields and open ground, taking vegetable leaves, beans, peas and grain.
Nests in holes in old trees, rocks, Rabbit burrows, buildings. Insubstantial structure of few twigs, bits of grass, or nothing at all.
Nestbox: Enclosed, with 8in (200mm) diameter entrance hole. 15in (380mm) interior depth, 15in × 25in (380mm × 635mm) floor. Provide a landing platform. May take to a tree-mounted Kestrel box.
Eggs: 2 creamy-white. Incubation 16–18 days; fledging 28 days. Three, four or even five broods.

**Wood Pigeon** (Ring Dove) *Columba palumbus*
Resident and generally distributed except in extreme north of Scotland. Open country of all kinds, provided there are some trees.
Feeds mainly on ground, but in spring will graze over foliage, buds and flowers in trees. Main food vegetable, cereals, roots, beans, peas and seeds. Will come to garden feeding station for ground food, bread, vegetable scraps, seeds, and may even visit the bird table. Partial to beans and peas.
Nests in tall hedgerows, almost any kind of tree, second-hand crow nests and squirrel dreys. Sometimes on ground or on building ledges in towns, where it has overcome shyness. Few twigs (you can often see the outline of the eggs if you stand under the flimsy nest).

Eggs: Normally 2 white eggs. April to September. Incubation 17 days; fledging about 3 weeks. Three broods usually.
Read: *The Wood Pigeon*, R. K. Murton, Collins, 1965.

**Collared Dove** *Streptopelia decaocto*
Resident and widely distributed. Vicinity of farm-buildings, park-like places and gardens in towns and villages. As recently as 1954 this species did not figure on the British List, yet it is now found throughout the country, the result of a remarkable cross-Europe invasion originating from India.
Finds its food in close relationship with Man, sharing grain with chickens, raiding corn and stackyards. In parks and gardens will also take berries and young foliage. Comes freely to a bird table and ground feeding station for seeds, peas, grain and scraps.
Nests in trees, preferably conifers, on a platform of sticks, grasses and roots.
Eggs: 2 white eggs. March–October. Several broods.

**Barn Owl** *Tyto alba*
Resident, generally distributed but not abundant and decreasing. Vicinity of farms, old buildings, church towers, etc. Parkland with old timber.
Hunts over fields and open country for small rodents and even small birds.
Nests in ruins or unoccupied buildings, hollow trees and cliff crevices. No material used, the eggs are often surrounded by a pile of cast pellets.
Nestbox: Barn Owls are a beneficial species from the point of view of farmers, hunting a diet of Short-tailed Voles, Common Shrews and Wood Mice. Formerly much persecuted by the ignorant, they have been further declining in this century because of habitat loss and human disturbance. But there is also a chronic shortage of suitable nest sites such as old trees, derelict buildings and old-style brick and timber barns. The modern steel-framed barns offer no home to nesting Barn Owls. Fortunately, they take readily to nestboxes, the bigger the better, particularly if the box is placed in a building which is not too often disturbed.
There are several designs, and it may be necessary to use a good deal of ingenuity in fitting the box to the site. The RSPB suggests that enclosed storage barns with access from outside are most favoured, but open Dutch barns are also suitable, particularly if the nestbox can be secured to beams or struts and used for roosting in winter. This will increase the chance of nesting the following year.
The design of the boxes is quite straightforward, as they can easily be made from a standard tea chest or packing case – both for lightness and ease of conversion they are hard to beat – but the bigger the better. If you can find a source of supply, wooden barrels are also easy to convert. Tea chests and packing cases, however, are not waterproof and should only be sited in dry locations. If the nestboxes are exposed to the elements, more durable and water-resistant material must be used. The lining papers and metal edging around the top of the chest must be removed and any nails knocked flat. You will also need some wooden trays – baker's trays are ideal – which are sawn in half to provide two platforms of about 18in (450mm) in depth. These are important as they provide a safe area in front of the box where the young can come out and stretch their wings.

## BARN OWL NESTBOXES

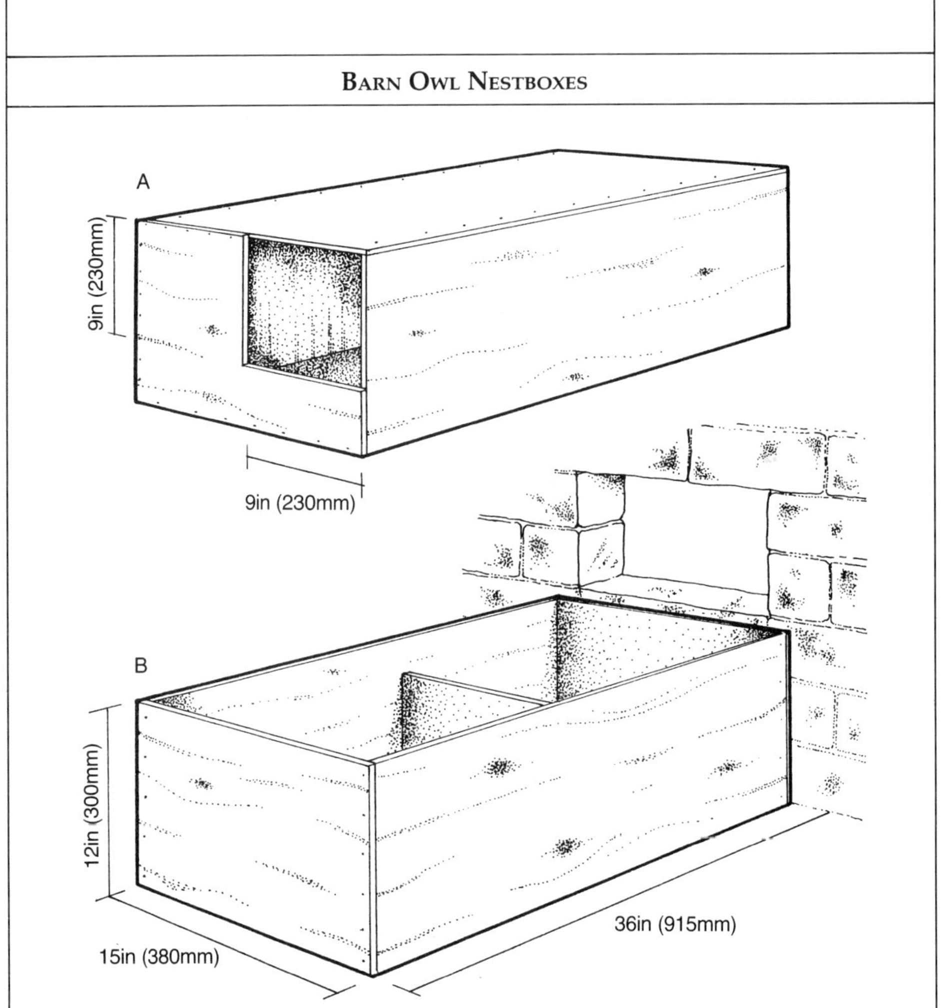


Barn Owl nestboxes. Type A should be used in timber-trussed or modern steel-framed barns. Type B is for enclosed storage barns with access from the outside.

It is easiest to erect a nestbox in a timber barn – a steel-trussed building may require considerable ingenuity. The most important points to remember are that the boxes must be secure inside the barn, as high above the ground as possible, in the darkest corner out of any draughts and where there is permanent access for the birds. Since height is one of the main criteria, the easiest time for putting up the boxes is when the barn is full of bales. In a timber-trussed barn the box is first nailed, from the inside, front and back to the beam with 3in (75mm) nails to give a firm fixing. The front, which has already had a 9in × 9in (230mm × 230mm) opening cut out of the corner, is then fixed to the open end with 1in (25mm) nails. Finally, the tray is nailed in front of the box. In some cases, it may be necessary to support the platform on timber runners nailed to the underside of the box.

If the barn has steel roof trusses, it is best to nail vertical and horizontal pieces of

timber to the box; these can be firmly roped, wired or G-clamped to the steelwork. Every ounce of ingenuity should be used when dealing with these barns, as very often they are the only suitable roosting and nesting-sites for miles around. Boxes can also be placed in corner sites and hung from ridge purlins, but virtually every barn demands its own solution. Barrels can be placed in disused lofts, but here access must be restricted while the birds are nesting.

Whether your boxes are occupied or not, keep the knowledge of their whereabouts restricted to as few as possible. Human predation is, unfortunately, a reality as is disturbance by well-meaning but misguided bird-watchers. Never let any unwanted eyes see you checking a building and only visit occasionally, preferably towards dusk, so that if the adult is inadvertently flushed, it will quickly return. Barn Owls are highly sensitive to disturbance. A most important point is that they are included on Schedule 1 of the Bird Protection Act. This means that both the bird and its eggs are specially protected by law, and if you intend to visit your occupied nestboxes you must obtain a special Government Permit. If you see that the box is occupied early in the breeding season, it is probably best to watch from a safe distance, thus avoiding disturbing the birds and the need to become involved in such legalities. The Bird Protection Laws do not hinder the farmer from going about his normal business using the barn! Clear accumulated debris of pellets every year or so. The Hawk Trust (for address see page 183) will help with advice.

Eggs: 4–7 white. March–July. Incubation 32–34 days; fledging about 10 weeks. Frequently two broods.

Read: *Owls*, John Sparks and Tony Soper, David & Charles, 1987; *The Barn Owl in the British Isles*, Colin Shawyer, Hawk Trust, 1987; *The Barn Owl*, D. S. Bunn, Poyser, 1982; *Owls*, Chris Mead, Whittet, 1987.

## A LITTLE OWL NESTBOX

Little Owl nestbox *from BTO News, Oct. 1976*

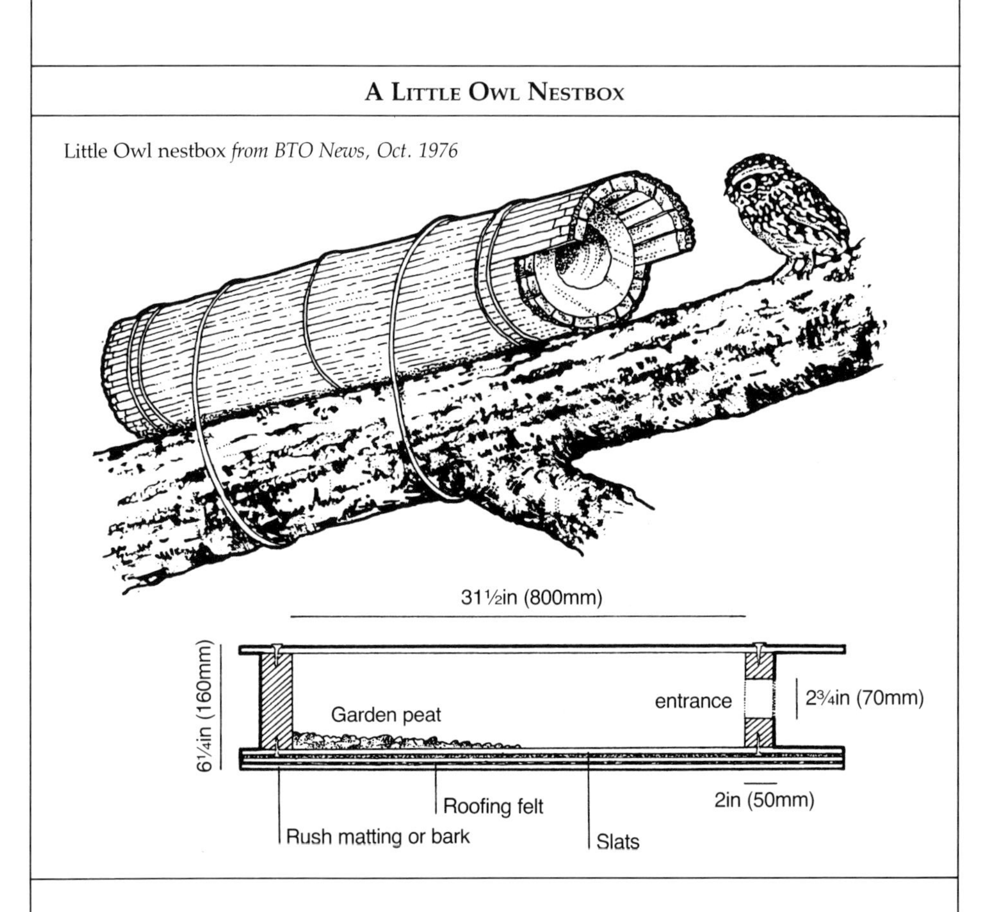


**Little Owl** *Athene noctua*
The Little Owl was first introduced to Britain from Italy by Charles Waterton in May 1842, though it was some thirty years later that a similar experiment was successful in the long term. Lack of suitable tree holes may be one of the reasons for its current decline. Resident in southern half of England. Hunts mainly at dusk and early morning for small mammals, insects and a few birds.
Nests in trees, farm-building holes and Rabbit burrows.
Nestbox: Enclosed, at least 4in (100mm) diameter entrance hole, inside depth 12in (300mm), floor 8in × 8in (200mm × 200mm). Also may use Kestrel box. One of the most successful designs is the 'hollow branch'. Take two round, equal-sized wooden discs of softwood, approximately 2in (50mm) thick and not less than 6in (150mm) diameter. Bore a 2¾in (70mm) entrance hole in one disc. Form a drum by nailing wooden slats 39in × 1in × ½in (1m × 25mm × 12.5mm) to the discs which should be 31½in (800mm) apart. Secure the drum with wire and wrap it with a layer of roofing felt. Camouflage with a layer of rush mat or loose bark and fasten with wire. Mount on a thick horizontal branch some 10ft–16ft (3m–5m) high, the drum sloping slightly to the rear. Don't expose the entrance to the prevailing wind. Prime with garden peat. It is important that the interior is not less than 31½in (800mm) long and the closed end must be light-proof.

Eggs: 3–5 white. April and May. Incubation 28–29 days; fledging about 26 days. Usually one brood.

**Tawny Owl** *Strix aluco*
Resident and generally distributed in Britain, but never recorded wild in Ireland. Woodland, farmland, parks and well timbered gardens.
Hunts at dusk for small mammals, birds and insects, even frogs and newts. May take scraps – and small birds – from bird table. May take Noctules or Pipistrelles in city centre.
Nests in tree holes, second-hand crow, hawk and heron nests, squirrel dreys. Sometimes in barns and on rocky ledges. Branch may need sawing above hole to prevent loss of nest site due to gales. Listen on a quiet night in January to map their territories and determine likely spots for a successful box.
Nestbox: Enclosed with 8in (200mm) diameter hole at top (see drawing), inside depth 30in (760mm), floor 8in × 8in (200mm × 200mm). Will use a barrel (40gal/180l best, but 6gal/27l has been used successfully) if a hole is opened in it and the barrel fixed to a tree crutch about 12ft–30ft (3.6m–9m) high, although the height is probably not critical. Or might use a Barn Owl box.
The chimney type nestbox (see drawing) has four wooden planks at least 30in (760mm) long and 8in (200mm) wide butted onto each other, using 2in (50mm) or so oval nails, to make a square-sectioned chimney. A 9in × 9in (230mm × 230mm) base, which must be perforated by at least half-a-dozen drainage holes, is nailed to one end to form the floor. A thin sheet of ferrous metal is to be preferred to either perforated zinc or a wooden floor. A layer of dry peat or sawdust should be added to the completed base to counteract the fouling that will occur in the fledging period. Chimney boxes of this size are too deep for a hand to reach to the

**TAWNY OWL NESTBOXES**

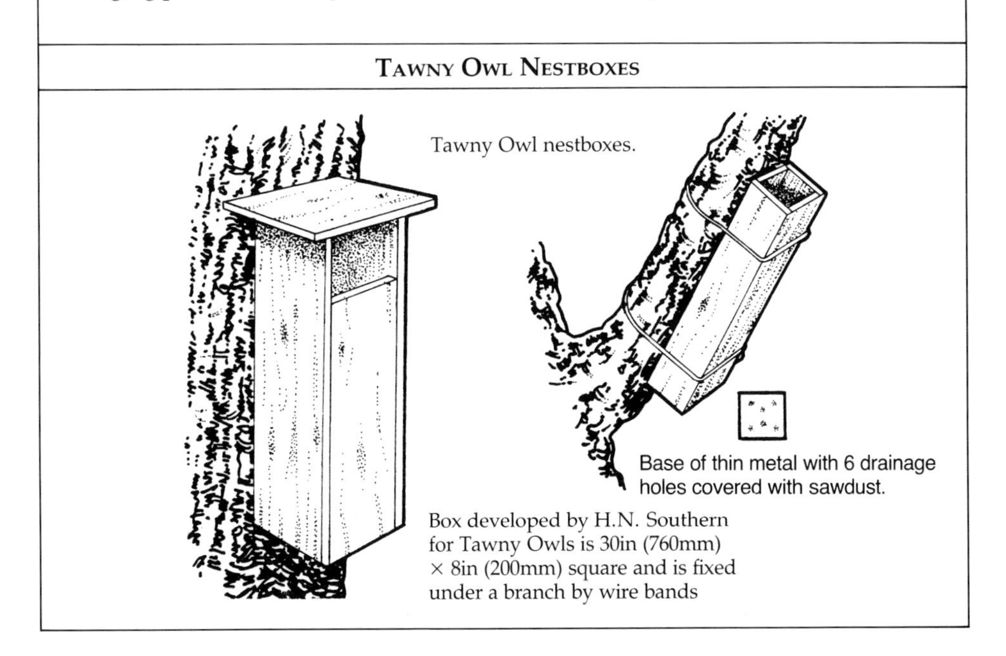


Tawny Owl nestboxes.

Box developed by H.N. Southern for Tawny Owls is 30in (760mm) × 8in (200mm) square and is fixed under a branch by wire bands

bottom, either for examining, or ringing the nestlings, or for cleaning out. It is practical to make an observation door on one side of the box 8in × 6in (200mm × 150mm), which is hinged to the back of the box and fastened at the front by a hook-and-eye catch. Fit the box under a lateral tree bough at an angle of about 30° from the vertical. If attaching to main trunk, contrive an angle of about 45° to simulate a broken branch. Secure to the tree by wire bands at both top and bottom, but remember these will rust through, or become embedded in bark, so watch your maintenance. These boxes have also been used by Kestrels, Robins, Great Tits, Jackdaws and Starlings. In Scotland, a box might attract a Pine Marten. Be cautious, Tawnies can be dangerous.
Eggs: 2–4 white. February to early April. Incubation 28–30 days; fledging about 4 weeks. One brood.

**Long-eared Owl** *Asio otus*
Locally distributed over most of British Isles, least common in south-west. Mainly in coniferous woods, plantations, shelter belts; also in well-ivied deciduous woods and marshes, dunes, moorland with low bushes.
Nests in second-hand crow, Sparrowhawk, pigeon or heron nests.
Nestbox: Has been known to use duck-type nest baskets in Holland. In the Cambridgeshire fens, David Garner has designed purpose-built willow baskets, 12in (300mm) diameter, 6in (150mm) deep and lined to a depth of 3in (750mm) with dead twigs, and placed them in hawthorn trees at a height of between 12ft–16ft (3.6m–4.8m). Sited late in the year, they have been readily accepted – but don't try them in Tawny Owl habitats, they can't compete.
Eggs: 4–5 white. March, early April. Incubation 27–28 days; fledging about 23 days. One brood.

Swifts return to this country in the middle of April and their screaming parties are a familiar sight and sound.

The Swift is superbly streamlined and adapted to a life on the wing.

Swifts nest in holes in buildings, usually inside roofs. They can be encouraged to use specially designed Swift nest boxes fixed under eaves.

## A Swift Box

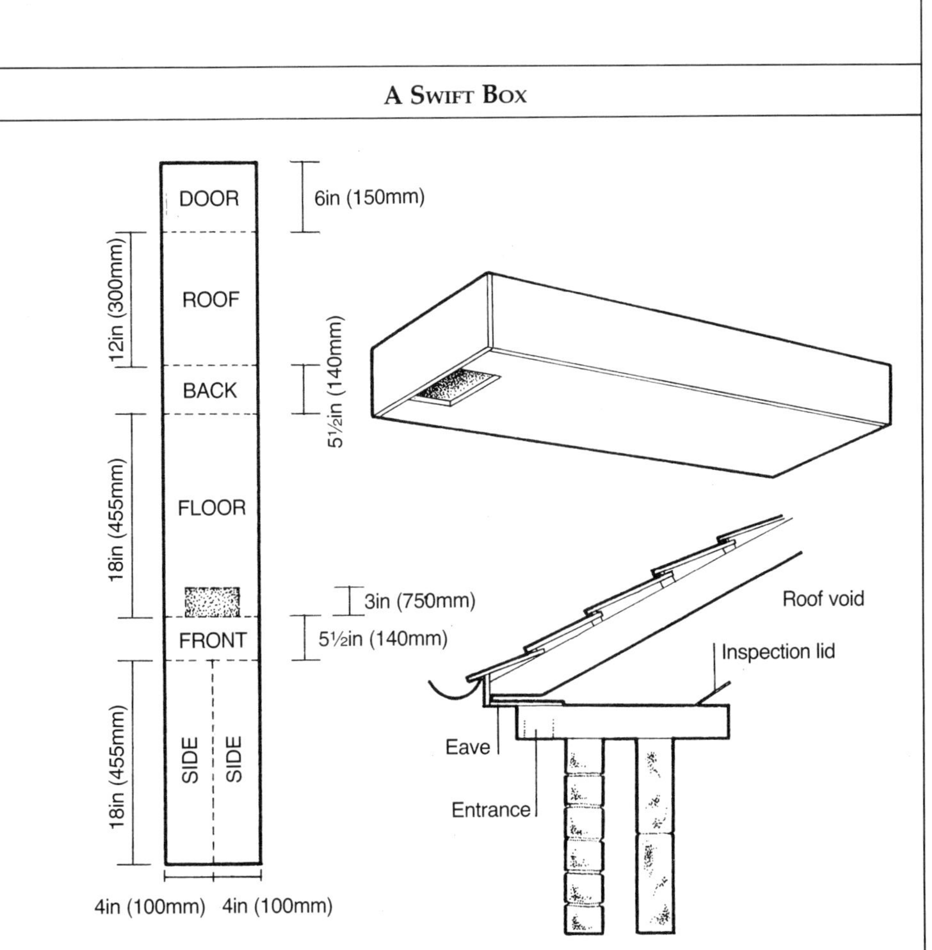


Swift nestbox in place on side of house. By removing a brick or section of wood from the eaves the box may be positioned within the loft with only the entrance hole visible from outside.

**Swift** *Apus apus*
Summer resident, generally distributed except in north-west Scotland, arriving late April, early May, leaving early August. Habitat exclusively aerial. Rarely on ground except at nest. Feeds on the wing, taking only insects, anywhere from ground level to 1,000ft.
Nests in colonies, under eaves, in crevices and in holes. Bits of straw, grass, feathers, seed fluff, collected on the wing and stuck together with saliva to form a cup.
Nestbox: Using a plank 65in × 8in × ¾in (1.65m × 200mm × 100mm), make a box 19½in × 8in × 5½in (495mm × 200mm × 140mm) with an entrance hole cut in the *floor* of the box (not the end, as Swifts prefer to enter vertically from below). Make box longer than 19½in (495mm) if convenient, but not shorter, as they like to nest at least 1ft (300mm) from the entrance hole. Prime nest area with a ring of

twisted straw. Cut an inspection door 6in × 8in (150mm × 200mm) at rear roof to aid cleaning. Site it under the eaves at least 12ft (3.6m) above ground, up to 100ft (30m) if necessary. Block entrance hole till Swifts first arrive, at the end of April or the beginning of May, in order to discourage earlier nesting sparrows and Starlings. Sensitive to disturbance.
Alternative plan: Open out a narrow slit in eaves to allow entrance to your roof.
Eggs: 2–3 white. Late May, early June. Incubation 18–19 days; fledging about 6 weeks. One brood.
Read: *Swifts in a Tower*, David Lack, Methuen, 1956.

**Kingfisher** *Alcedo atthis*
Resident and generally distributed, except in Scotland. Streams, rivers, canals, lakes, estuaries (especially in winter). Perches or hovers above water, fish-watching. Plunges to capture small fish, insects, larvae and amphibians. In winter, visits coast for shrimps, prawns, small rock-pool fish, etc. May come to garden ponds for Minnows or Sticklebacks. Nests in tunnels in banks of streams or sand pits, boring 2in (50mm) tunnels as far as 4ft (1.2m) to a nest chamber, preferably in sandy soil.
Artificial nest-site: Bore a 15° upward sloping tunnel into an uncluttered vertical or near-vertical north-east facing stream bank. At least 3½ft (1.05m) of bare face is needed to discourage predators. The tunnel should be 4in (100mm) wide and 2½in (60mm) high and at least 3ft (900mm) or so above high-water mark. Leave birds to excavate nest-chamber or use ingenuity to create a chamber 7in (175mm) round and 5in (127mm) high.
Or, will excavate tunnel if an artificial bank is provided beside a suitable stream, a feature developed by Ron and Rose Eastman. Fix fencing posts eg willow, which will sprout even if embedded in concrete. Stretch 2½in (63mm) mesh chicken wire or square mesh pig wire, to a height of 4ft (1.2m) or more to make a vertical face, facing north if possible, with a degree of privacy and foliage. Fill in behind wire wall with sand or sandy soil. Provide perches and posts nearby. There is no need to cut entrance hole in the netting. The Wildfowl Trust at Arundel in Sussex have had success with this design.
Eggs: 6–7 white. Late April to August. Incubation 19–21 days; fledging 23–27 days. Two, three or even more broods.
Read: *The Kingfisher*, D. Boag, Blandford, 1988.

**Hoopoe** *Upupa epops*
Passage migrant, regular in small numbers in spring, less frequent in autumn, on south, south-east and south-west coasts and in east coast as far north as Norfolk. Rare elsewhere in Great Britain. Open woodland, orchards, parkland.
Feeds mainly on ground, often close to human habitation, probing on lawns for insect larvae, etc. Does not often come to bird station, but might do so if mealworms/caterpillars/ant pupae were made available in dish. Not shy.
Normally breeds in Eurasia, but occasionally a Hoopoe will nest in one of the southern coastal counties, choosing tree holes, crevices and holes in rough stone walls and ruins.
walls and ruins.

Nestbox: Uses them on the continent, presumably may do so here. Enclosed, with 2½in (60mm) entrance hole, interior depth 10in (250mm), floor 6in × 6in (150mm × 150mm).
Eggs: 5–8 whitish-grey or yellowish-olive. May and June. Incubation 18 days; fledging 20–27 days. Two broods.

**Wryneck** *Jynx torquilla*
Summer resident. Deçreasing and scarce in south-east England with a very few pairs now left. However, there are signs of influx to Scotland from Scandinavia, with birds breeding in the Spey Valley, for example.
Picks up insects from tree surface, clinging to trunk like woodpecker. Sometimes on ground, picking up ants, insect larvae. Feeds occasionally at bird tables.
Nestbox: Improve a tree hole. Or try an enclosed box, with a ⅜in × 1¾in (10mm × 45mm) diameter entrance, a 6in (150mm) interior depth, and a 5in × 5in (127mm × 127mm) floor. No need for priming material.
Eggs: 7–10 white. End of May till July. Incubation 12 days; fledging 19–21 days. Usually one brood.

**Green Woodpecker** *Picus viridis*
Resident but local in England and Wales, rare in Scotland, none in Ireland. Deciduous woods, parks and farmland.
Searches for insect larvae over tree trunks and branches, probing with long mobile tongue; also feeds freely on ground, especially where there are ants' nests. In times of hard frost, when ant hills are frozen solid, it may damage beehives by boring holes to reach the insects within. May also attack nestboxes. Will visit bird table for mealworms, bird pudding, etc.
Nests in tree trunks, choosing soft or rotting timber, boring a hole horizontally 2in–3in (50mm–75mm) then descending to make a nest compartment over 1ft (300mm) deep and about 6in (150mm) wide at its broadest. Put a few chips at the bottom to form the nest. Sometimes, old holes are used again. Often, Starlings take over from them.
Nestbox: Enclosed type with 2½in (63mm) entrance hole, interior depth 15in (380mm), floor 5in × 5in (127mm × 127mm). Fill with polysytrene chips. Empty box ideal for Starlings!
Eggs: 5–7 translucent. End of April to May. Incubation 18–19 days; fledging 18–21 days. One brood.
Read: *My Year with the Woodpeckers*, Heinz Sielmann, Barrie & Rockliff, 1959.

**Great Spotted Woodpecker** *Dendrocopus major*
Resident, widely distributed in England, central and southern Scotland, none in Ireland. Wooded country – coniferous in north, deciduous in south – hedgerows, orchards and large gardens.
Hunts over trees for insect larvae, spiders, seeds and nuts, even wedging a nut into a tree crack to deal with it. Will come to bird table for suet especially; also oats, nuts, boiled fat bacon, hanging fat, or nuts. As adept as tits at feeding upside down.

Great Spotted Woodpeckers can become enthusiastic bird table customers; their Lesser Spotted relations are more elusive.

Nests in tree holes 10ft (3m) and higher from ground. Few wood chips form nest.
Nestbox: Enclosed type, entrance hole 2in (50mm), interior depth 12in (300mm), floor 5in × 5in (127mm × 127mm). Fill with polystyrene chips.
Eggs: 4–7 white eggs. May to June. Incubation 16 days; fledging 18–21 days. One brood.

**Lesser-Spotted Woodpecker** *Dendrocopus minor*
Resident in southern England and Midlands, becoming local and rarer further north. Widely distributed but scattered in Wales. Not in Scotland or Ireland. Same type of country as Greater Spotted.
Elusive bird, searching upper parts of trees for insect larvae. Will come somewhat nervously and rarely to bird table for fats, nuts and fruit.
Bores nest-hole in decayed soft wood of branch or tree trunk.
Nestbox: Enclosed, entrance hole 1¼in (30mm), interior depth 12in (300mm), floor 5in × 5in (127mm × 127mm). Fill with polystyrene chips.
Eggs: 4–6 translucent eggs, early May to mid-June. Incubation 14 days; fledging 21 days. One brood.

**Skylark** *Alauda arvensis*
Has been known to feed regularly on bread-and-cheese scraps from a garden of waste ground at the edge of Liverpool. In hard weather may come to seeds at a ground feeding-station. In 1977 a pair of Skylarks bred on the grassy turf roof of the visitor centre at the Wildfowl and Wetlands Trust reserve, Martin Mere, in Lancashire. By 1985 seven pairs had taken up residence on this high-level and Fox-free site.

**Sand Martin** *Riparia riparia*
Summer resident, widely distributed. Open country with water.
Feeds mainly over water, taking insects on the wing. Perches on wires and low branches, and will occasionally pick insects from the ground while on the wing.
Nests colonially, digging a long tunnel to a nest chamber, in sand and gravel pits,

Swallows return in early April having spent the winter in Africa. The male's tail streamers are longer than those of the female.

Juveniles do not have long tail streamers.

Swallows typically hunt for flying insects at low level over fields and water where they will also drink and bathe on the wing.

Swallows nest under cover on ledges inside buildings or in porches, especially around farms.

House martins return from winter quarters before the end of March.

The enclosed cup-shaped nest is made with pellets of mud collected from the edges of ponds or puddles.

In Autumn, prior to migration, large numbers of House Martins congregate with Swallows and Sand Martins on wires and roofs.

Sand Martins arrive in mid March and are usually first seen over stretches of water.

In Autumn huge numbers gather to roost in beds of reeds and osiers.

Sand Martins nest colonially in sand pits, quarries, cliffs and places where they can excavate 2–3 foot long nesting tunnels.

RSPB

A SANDMARTIN NESTBOX

Sand Martin nestbox.

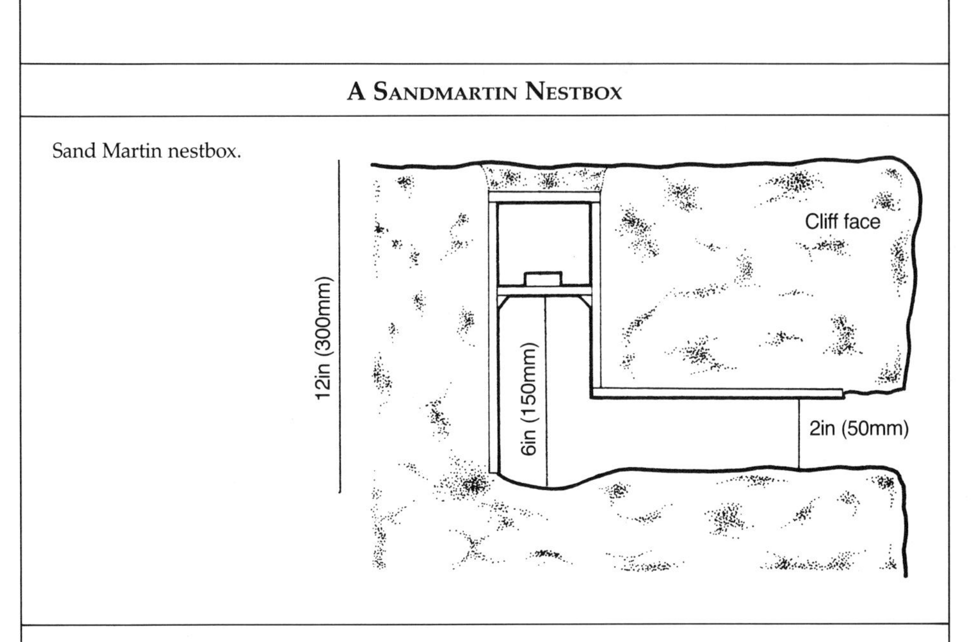


railway cuttings, river banks and sea cliffs. Few grasses and feathers. Try boring a few enticing 2in (50mm) diameter holes in likely sandbanks, steep road cuttings and banks, especially over water – they may use drainage pipes in a wall which can be deliberately placed for them. A few score breed in this sort of location at the RSPB's Minsmere Reserve in Suffolk, at the car park.

Nestboxes: Near an existing colony it is worth preparing some underground chambers. Dig a vertical shaft 1ft (300mm) deep, line it with boxing, arrange an access tunnel 2in (50mm) square, from the vertical cliff entrance. The length of the horizontal passage is not critical. Close the nest chamber with a removable lid 6in (150mm) from the floor. Close the roof of the shaft with another lid which is concealed by a turf.

Eggs: 4–5 white. Mid May onwards. Incubation 14 days; fledging 19 days. Two broods.

**Swallow** *Hirundo rustica*

Summer resident, generally distributed. Open farmland, meadows, ponds. Spends much time in flight, especially over water, hunting insects from ground level to 500ft. Unlike Swifts, settles freely on buildings and wires. Seldom on ground, except when collecting mud for nest. (Provide mud puddles in times of drought.)

Nests on rafts and joists, building open mud-and-straw cup, lined with grasses and feathers.

Nestbox: Improvise a simple saucer shape, or fix half a coconut or a 4in × 4in (100mm × 100mm) shallow tray to joist or rafter, even as low as 6ft (1.8m). Will also use specially adapted House Martin nestbox placed singly inside building. Or, using an old nest, make a plaster of Paris mould of the interior. Then taking potting clay to make a thick replica, complete with fixing flanges or saddles to fit

over a joist, remembering that Swallows like to nest *against* something. Remember to allow continuous access to the nest-site.
Eggs: 4–5 white, spotted with red-brown. Mid May till October. Incubation 15 days; fledging 3 weeks. Usually two broods.
Read: *A handbook to the Swallows and Martins of The World*, Angela Turner, Helm, 1990.

**House Martin** *Delichon urbica*
Summer resident, generally distributed.
Habitat as Swallow, but more often near human habitation.
Hunts insects on the wing, especially over water. Also on the ground.
Originally a cliff nester, has now adopted buildings. Nests colonially on outside walls, under eaves. Cup shape made of mud gobbets with feathers. In dry summer, provide mud puddles for building materials of different consistencies.
Nestbox: Artificial nest from Nerine Nurseries (for address see page 185). Fix under eaves or high window sill. For best results an existing House Martin colony should be close at hand. There is some evidence that House Martins prefer to nest on north and east facing walls. One nest may work, but the more the merrier. Put them in groups outside, under the horizontal or sloping eaves of houses, barns, etc. The artificial cups are held in position by cup-hooks so that it is possible to slide the nest freely in and out to inspect the contents. The entrance hole for House Martin nest cups should be no more than 1in (25mm) deep, in order to exclude sparrows. Nevertheless, there have been cases where the hole has been enlarged and sparrows have gained access. There is a method which has been successful in stopping this, based on the fact that House Martins are able to approach a nest at a much steeper angle than sparrows (see drawing below). The hanging cords of the curtain should be no more than 12in (300mm) long and should be fixed to hang 6in (150mm) away from the entrance hole. A spacing of 2½in (63mm) between the cords is effective. Use ⅞in (22mm) steel nuts as weights on the cord ends. Have all the cords the same length so that they are less likely to tangle in a wind.

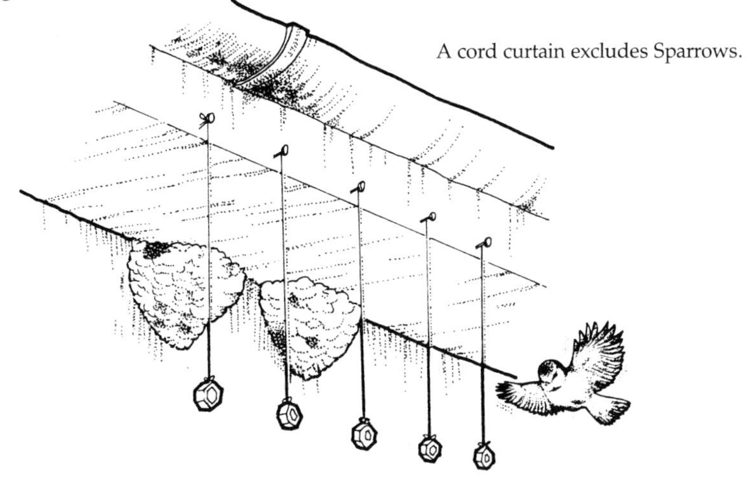
A cord curtain excludes Sparrows.

One of the objects of using artificial nests is that they frequently encourage House Martins to adopt a house not previously 'tenanted' and make their own nests. So even if the boxes are not used, they may be successful in their purpose. But single nests away from an existing colony are susceptible to attack from sparrows. If the birds try to build and the nests fall off the eaves, a series of nails in the facia board may help with their adhesion.

It is possible to construct your own artificial nest using Polyfilla or a mixture of cement and sawdust. As a model you can use either a plaster of Paris mould of an old nest or a quarter segment of a plastic ball about 7in (175mm) in diameter. The Polyfilla or cement mixture should be smoothed over the model to a thickness of about ⅓in (8mm), leaving a flange around the edge to facilitate fixing. The hole should be cut no deeper than 1in (25mm) and no more than 3in–4in (75mm–100mm) wide. The nest can either be mounted on a board and then fixed up with cup hooks, which allows for easy removal, or it can be fixed directly to the house using Polyfilla. It is often a good idea to smear the outside of the nests with mud, especially around the hole.

Rearing young House Martins: Occasionally a House Martin's nest may fall with the young still inside it. The use of a substitute nest may encourage the parents to continue feeding them. A strong box, deep enough so that they cannot fall out and replaced near the original nest site, is usually successful. In one case an old Blackbird's nest was used and the parents rebuilt around it. If the parents have deserted, or it is not possible to use a substitute nest, then the young must be fed by hand, and if the birds will not feed by themselves they will have to be force-fed using a pair of forceps.

It will be difficult to catch enough insects with which to feed a hungry young House Martin, but there are a few substitutes. Hard-boiled egg yolk mixed with crushed soaked biscuits provides the necessary nutrients and is similar to the proprietary insectivorous bird food available from pet shops. Cut mealworms and maggots make a good food but a little soaked bread must occasionally be given in order to provide enough calcium. The birds can be fed every two hours, giving them 4–6 maggots or a similar amount of egg mixture.

The birds should be housed in a suitably sized box and kept in a warm place. When ready to fly the best place for them is an aviary where they can practise flying and feeding before release. Fledged birds can be encouraged to feed themselves by threading mealworms of maggots with cotton and suspending them from the ceiling. Care must be taken that the birds do not get tangled in the cotton.

One of the disadvantages of having House Martins nesting on your house (especially over a doorway) is the droppings that will fall beneath the nest. The simplest way to overcome this problem is to fix up a shelf 10in (250mm) wide about 6ft (1.8m) below the nest, which should catch any droppings. A removable shelf can be made using keyhole brackets. The Wildlife and Countryside Act 1981 now makes it an offence intentionally to take, damage or destroy the nest of a House Martin while the nest is in use or being built.

Eggs: 4–5 white. Late May to October. Incubation 14–15 days; fledging 19–21 days. Usually two broods, often three.

**Meadow Pipit** *Anthus pratensis*
Abundant. Open country, moors, heaths, sand dunes, wintering on lowland pastures, sewage-farms, sea coast.
Feeds mainly on insects, some weeds. Early winter is the most likely time to see it in the garden.
May come to ground feeding stations, eg joining Dunnocks under a bird table, especially near traditional wintering places.

**Rock Pipit** *Anthus spinoletta*
Resident and generally distributed. Rocky shores in summer; marshes, waterways, estuaries and coasts in winter.
Forages near water for insects, animals and vegetable matter. Will come freely to a ground feeding station near the shore for crumbs, scraps, cheese.
Nests in hole or cliff-crevice close to the shore. Stems and grasses, lined with grass and hair. May take to a fish-box, placed upside down and with a 2½in (50mm) square entrance hole in one end and another similar hole in one of the long sides. Put a heavy stone on top to secure it.
Eggs: 4–5 greyish-white eggs spotted olive-brown and ashy-grey. Late April to June. Incubation about 14 days; fledging 16 days. Two broods.

**Grey Wagtail** *Motacilla cinerea*
Ledge or open-fronted box, as for Pied Wagtail, but placed low over a suitable stream and underneath an overhang, for instance a bridge.

**Pied Wagtail** *Motacilla alba*
Resident and generally distributed. Gardens, farms, buildings and cultivated country.
Restless bird, feeds over ground, but often flutters up to take an insect. Mainly insects, fond of shallow pool edges. Will come freely to ground feeding station, scavenging crumbs and scraps where other birds have left unconsidered trifles.
Nests in holes and on ledges of walls, outhouses, creeper, banks and cliffs. Leaves, twigs, stems, lined with hair, wool and feathers.
Nestbox: Ledge or open-fronted box, with a floor area of not less than 4in × 4in (100mm × 100mm). Fix it in a stone wall. Or make a cavity behind a loose stone which can be used as an inspection door.
Eggs: 5–6 greyish- or bluish-white, spotted grey-brown and grey. Late April to June. Incubation 13–14 days; fledging 14–15 days. Two broods. Often host to cuckoo.

**Waxwing** *Bombycilla garrulus*
Irregular winter visitor, usually to eastern counties.
Food: Hedgerow berries. Cotoneaster berries in gardens. May come to bird table for fruit in hard weather.

**Dipper** *Cinclus cinclus*
Resident, generally distributed in suitable localities. Fast-flowing streams and rivers of hills and mountainous regions.
Food: Aquatic insects.
Nests in wall and bridge holes, rock faces, tree roots and under waterfalls, always close to fast moving water. Construction of mosses, grasses under an overhang.
Nestbox: May occasionally occupy an open-fronted Robin-type box. A German design has been developed to provide nest recesses in the supports of concrete bridges at the time of construction. The essence of the operation is that a recess (see drawing) is left in the concrete by making a mould which is filled with expanded polystyrene and then inserted into the mould for the concrete bridge section. Once the concrete has set, the entrance hole is chipped out and the

NEST RECESS FOR DIPPERS

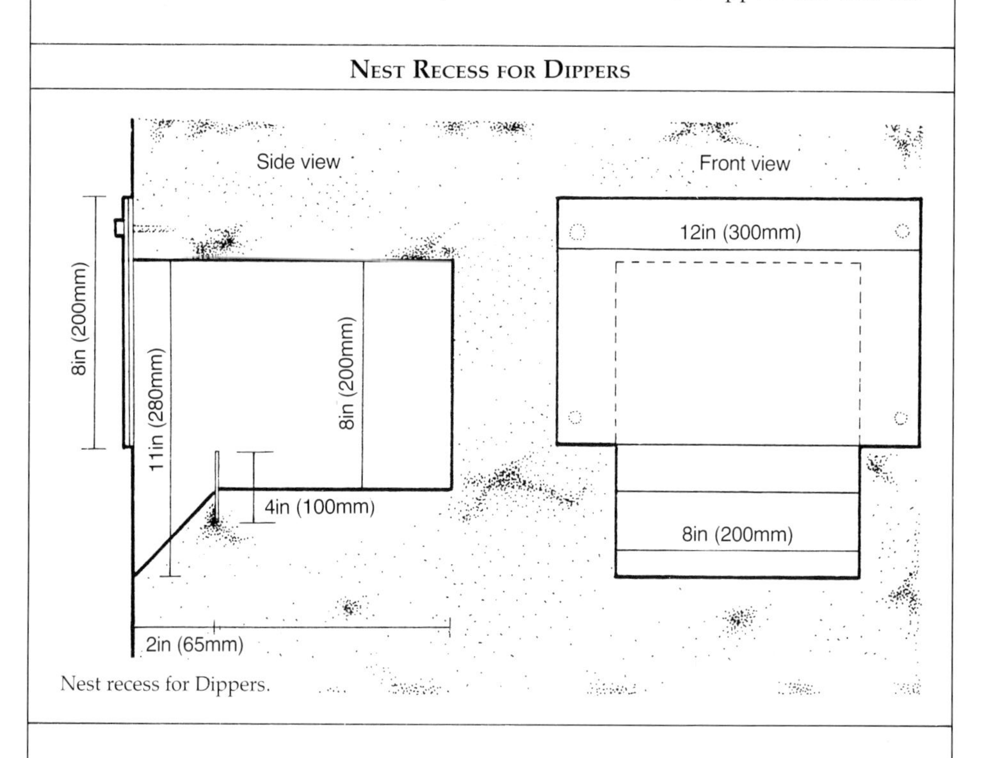

Nest recess for Dippers.

polystyrene removed. The only remaining task is to fit a front section over the opening. Arrange a suitable perch or dipping stone just above water level if there isn't one already.
Eggs: About 5 white. End of March. Incubation about 16 days; fledging 19–25 days. Usually two broods.

**Wren** *Troglodytes troglodytes*
Resident and generally distributed. Gardens, thickets, woods, rock banks. Avoids the centres of large towns.
Lives in a world of cracks and crevices, twigs and woodpiles, hedgebottoms, and the mysterious undergrowth round fallen trees. Active and diligent hunter for insects and spiders. Will take crumbs, but is not a common bird table visitor.
Nests in hedges, holes in trees, banks or buildings. Cock bird makes several nests of moss, grass, leaves, etc, and the hen lines her choice with feathers sometimes weeks after the male built it.
Nestbox: May take to a tit box, but is much more likely to find a natural or semi-natural place such as a faggot pile or creeper-clad wall. Excavate a cavity in a bundle of pea sticks or brushwood and lean it against a wall. Provide a coil of rope in the corner of a shed, or hang up an old coat with capacious pocket.
Eggs: 5–6 white, spotted with brownish red. Incubation 14–15 days; fledging 16–17 days. Usually two broods.
Read: *The Wren*, Edward A. Armstrong, Collins, 1955.

**Dunnock** (Hedge Sparrow) *Prunella modularis*
Resident and generally distributed. Gardens, shrubberies, hedgerows.
Forages unobtrusively on ground among dead leaves, hedgerow bottoms, etc. Weed seeds in winter, insects in summer. Will come freely to ground station, less readily to bird table for crumbs of cornflakes, cake, biscuit, seeds. Unlike most other birds will eat lentils.
Nests in hedges and evergreens, faggot heaps. Twigs, moss, leaves, etc, lined with moss, hair and feathers.
Eggs: 4–5 deep blue eggs. April onward. Incubation 12 days; fledging 12 days. Two broods. Often serves as host to Cuckoo.

**Robin** *Erithacus rubecula*
Resident and generally distributed, except in extreme north of Scotland. Gardens, hedgerows, woods with undergrowth.
Feeds freely in open and in undergrowth. Insects, spiders, worms, weed seeds, fruit, berries. Has a flattering relationship with Man and will follow the digging spade hopefully.
Enthusiastic bird-tabler, very fond of mealworms, will also take seeds, nuts, oats, pudding, etc. Fond of butter and margarine, and is alleged to be able to tell the difference! Unsociable bird, it will endure the close company of its relations at the bird table only in hungry times.
Nests in gardens and hedgerows in bankside hollows, tree holes, walls, amongst creeper, on shelves in outbuildings, often at foot of bush or grassy tuft. Foundation of dried leaves and moss, neatly lined with hair and perhaps a feather or two.
Nestbox: Ledge or tray, open-fronted box. Interior floor at least 4in × 4in (100mm × 100mm). Old tin, watering can, or kettle, at least quart-size, well shaded from sun, spout down for drainage. Fix it about 5ft (1.5m) up in a strong fork site. Prime with a plaited circle of straw.

Eggs: Usually 5–6 white, with sandy or reddish freckles. Late March to July. Incubation 13–14 days; fledging 12–14 days. Two or more broods.
Read: *The Life of the Robin*, David Lack, Witherby, 1985; *Robins*, Chris Mead, Whittet, 1984.

**Black Redstart** *Phoenicurus ochruros*
Passage-migrant and winter visitor, some staying to breed in southern England. Cliffs, large old buildings, industrial premises, rocky and 'waste' ground such as building-sites, dumps or ruins.
Restless bird, in trees, over buildings, and on ground. Hawks for insects. Also takes berries. Although it lives amongst us, it is independent of our food, yet surely it might come to a feeding station for minced meat or berries, especially in hard weather.
Nests in crevices and holes of rocks or buildings, or on rafters, under eaves in outbuildings. Loosely-made of grass, moss, fibre, etc, lined with hair and feathers.
Eggs: 4–6 white eggs. Early April onward. Incubation 12–13 days; fledging 16–18 days. Two broods.

**Redstart** *Phoenicurus phoenicurus*
Summer resident, widely distributed but local. Woodlands, parks, bushy commons with old trees, ruins, orchards, well-timbered gardens.
Restless bird, flitting amongst branches or hawking for insects. Might come to bird table for berries, fruit and mince meat. Was a regular visitor to canteen door at Millwall Dock in East London in severe weather for grated cheese, mashed potato and cat food (after the cat had finished).

A REDSTART BOX

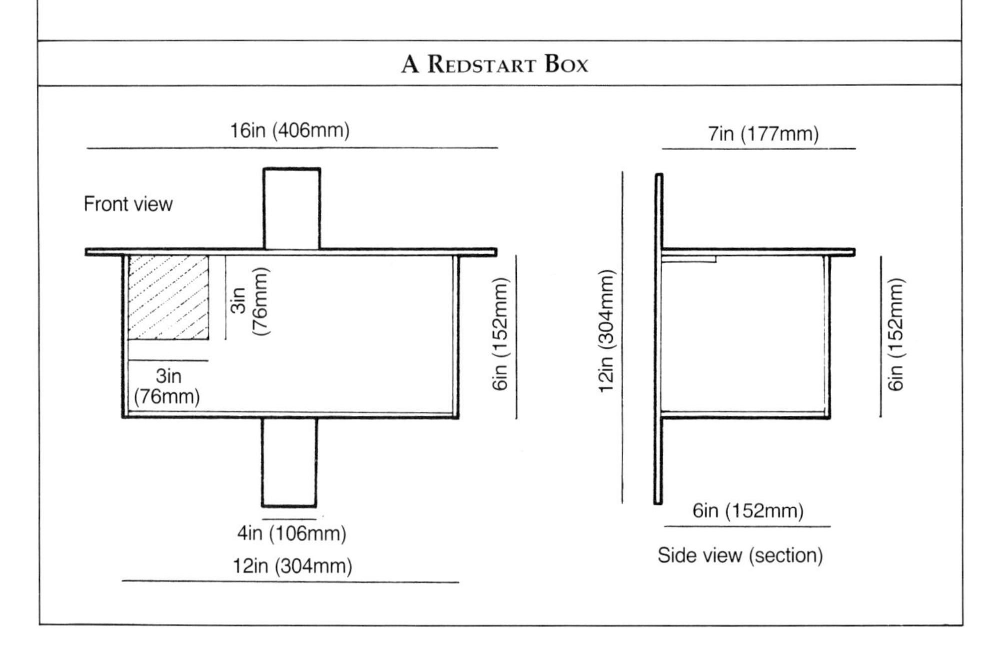

Nests in holes such as tree or stump, buildings, walls, outhouses, rocks, quarries. Nest made of grass, strips of bark, mosses, roots, and lined with hair and feathers.
Nestbox: Enclosed with entrance hole 1⅛in–2in (29mm–50mm) diameter, inside depth not less than 5in (125mm), and floor not less than 4in × 4in (100mm × 100mm). Make sure there is a perch not far away from the box (but not on it). Or try the Pete Jennings design shown in the drawing (this page).

Eggs: About 6 pale blue. May onwards. Incubation 14 days; fledging 14 days. Sometimes two broods.

**Wheatear** *Oenanthe oenanthe*
Summer visitor, locally common in open country, loose boulders and scree, Rabbit warrens, moorland, highland roads, chalk downs, sandy commons and sand dunes, stony shores, rocky islands.
Food: Insects.
Nests under a boulder or stone, in Rabbit burrow or stone wall. Grass and moss, linings of Rabbit fur, feathers, wool.
Nestbox: Has nested under tin cans and in tunnel-type prefabs made of bricks sunk in the shingle at Dungenness Bird Observatory. Be careful to protect the chamber from excessive sunshine, by piling soil or gravel over any metal parts. Likes a few strategically placed twig perches on the flight path to box.
Eggs: Usually 6 pale blue. Late April or May. Incubation 14 days; fledging about 15 days. One brood.

**Blackbird** *Turdus merula*
Resident and generally distributed. Commonly found in woods, hedges, gardens, shrubberies.
Feeds in open and in undergrowth, but never far from cover. Makes surprising amount of noise as it searches among dead leaves for insects, worms (which it often steals from a Song Thrush), fruit, berries and seeds. Will come freely to ground station and to bird table, for sultanas especially, cheese, fat, apples, cake, Rice Krispies, berries, seeds.
Nests in hedges, bushes, evergreens, Ivy, sometimes in outhouses. Sturdily built of grasses, roots, etc. Inner mud cup lined with grasses.
Nestbox: Tray or open-fronted box, with a floor area 12in × 12in (300mm × 300mm). Or try an inverted cone made of roofing felt. Cut into circle 9in (230mm) in diameter. Cut out and reject a V-shaped sector from centre to a 2in (50mm) arc at periphery. Cut a 1in (25mm) section from centre (to provide drainage). Now overlap open ends 3in (75mm) and staple strongly. Resulting cone is approx 7in (175mm) in diameter with a depth of 2in (50mm). Or try a bundle of pea sticks arranged with a central cave.
Eggs: 4–5 bluish-green, freckled with red-brown. February (or even earlier) to July. Incubation 13–14 days; fledging 13–14 days. Two or three broods, the first often being vandalised because it is insufficiently concealed by leaves, later broods being more successful. Five broods have been raised in one season.
Read: *A Study of Blackbirds*, D. W. Snow, Natural History Museum, London, 1988.

**Fieldfare** *Turdus pilaris*
Winter visitor, generally distributed. Open country, field and hedges.
Flocks feed in open formation across fields, looking for slugs, spiders, insects. In hedgerows, for berries of hawthorn, Holly and Rowan, yew, etc. In hard weather will come to ground station or bird table for berries, fruit, seeds, pudding, etc.
Breeds in Scandinavia, Central and Eastern Europe and Siberia.

**Song Thrush** *Turdus ericetorum*
Resident and generally distributed. Parks, woods, hedges, shrubberies and gardens, especially around human habitation.
Forages in open and in undergrowth for worms, slugs and especially snails, which it smashes on 'anvil' stones. Also insects, windfalls, berries and seeds. Will eat soft fruit, but is beneficial on the whole. Somewhat nervous visitor to ground station, not so often on bird table; fond of sultanas, also currants, cheese, fat, apples and scraps.
Nests in hedgerows, bushes, trees, among Ivy, occasionally in buildings. Strongly built of grasses, roots, etc. Stiffened with mud and with a unique lining of rotten wood or dung mixed with saliva and moulded into shape by the hen's breast. If they need protection from Magpies, contrive a canopy of chicken wire over nest.
Eggs: 4–5 eggs, blue with greenish tinge, spotted black or red-brown. March to August. Incubation 13–14 days; fledging 13–14 days. Two or three broods.

**Redwing** *Turdus musicus*
Winter visitor, generally distributed. Open country and open woods.
Feeds in loose flocks in fields or woods. Worms, slugs, snails, insects. Hawthorn, Holly, Rowan, yew berries. In hard weather will come to ground station or bird table for berries, seeds, scraps, fruit, etc.
Breeds in Scandinavia, Eastern Europe and Siberia.

**Mistle Thrush** (Stormcock) *Turdus viscivorus*
Resident and generally distributed, except in high mountains and treeless districts. Large gardens, orchards, woods.
Feeds mainly on ground, although it likes to sing from the highest point of a tree. Thrives on berries and fruit without conflicting with gardeners' interests. Yew and Rowan especially, but also hawthorn, Holly, Mistletoe, juniper, rose and Ivy. All wild fruits except blackberry. Will come to ground station, less freely to bird table for sultanas, currants, and bird pudding. Partial to bird garden life without being particularly friendly to Man.
Nests usually in tree fork or on bough. Grasses, moss, etc. Strengthened with earth and lined with fine grasses, the rim ornamented with lichens, bits of wool, feathers, etc.
Eggs: 4 tawny-cream to greenish-blue eggs, splotched with brown and lilac. February to April. Incubation 13–14 days; fledging 14–16 days. Frequently two broods.
Read: *British Thrushes*, Eric Simms, Collins, 1978.

**Blackcap** *Sylvia atricapilla*
Summer resident, frequently winters, local but fairly distributed, except in remote north and west. Open woodland, thickly bushy places, gardens with trees.
Active bird, searching in cover for insects, fruit and berries. Not often on ground. Will come to bird table for a wide range of food, including rolled oats, berries, crumbs and scraps, especially in hard weather. Overwintering birds are seen mostly near south coast, south-west peninsula and southern Ireland. Windfall

apples are an important food item, but they also take Cotoneaster, Honeysuckle, and Holly berries – Ivy berries as a last resort. After a heavy snowfall they may take Mistletoe berries, rubbing the berry in order to extract the seed. Apart from thrushes, few birds seem interested in Mistletoe berries.
Nests in bushes (especially Snowberry), hedgerows, evergreens. Stems, roots and grasses, lined with finer grass and hair.
Eggs: 5 eggs, light buff or stone ground, blotched brown and ashy. Mid-May onward. Incubation 10–11 days; fledging 10–13 days. Often two broods.

**Chiffchaff** *Phylloscopus collybita*
Summer resident.
Insectivorous bird, may come to bird table for kitchen scraps. Overwintering birds more likely to visit bird table.

**Goldcrest** *Regulus regulus*
Resident and generally distributed except in remote north-west. Woods, coniferous gardens, hedgerows.
Active and tame bird, which flits from twig to twig searching for spiders and insects. Will come to bird table and to hanging fat.
Nests in thick foliage of conifer. Ball of moss lined with feathers, held together by spiders' webs and suspended from branch.
Eggs: 7–10 white/ochreous eggs, spotted brown. End April to early June. Incubation about 16 days; fledging about 18–20 days. Two broods.

**Firecrest** *Regulus ignicapillus*
Few resident in New Forest, scarce passage migrant along south coast. Woods, gardens, scrub, bracken.
Habits more or less as Goldcrest. Will occasionally take suet or fat from crevices in tree-bark, etc. Has taken bread crumbs from bird table.
Nests in tree- or wall-holes; a loose nest of leaves, bark and mosses, lined with hair and feathers.
Nestbox: May use tit box.
Eggs: 5–9 white, speckled-brown eggs, late May. Incubation 12–13 days; fledging about 13 days. Single brood.

**Spotted Flycatcher** *Muscicapa striata*
Summer resident, generally distributed. Gardens, parks, woodland edges.
Sits on an exposed perch, flits out frequently to hawk after flying insects.
Nests against wall or on small ledge supported by creeper or fruit trees, etc. Moss and grass, lined with wool, hair or feathers.
Nestbox: Ledge or open-fronted box with at least 3in × 3in (75mm × 75mm) floor. Likes a clear view, so front wall only 1in (25mm) high, enough to retain nest. Hide one of those bowl-shaped wire flower baskets in dense Honeysuckle, primed with some moss. Have a perch not far away.
Eggs: 4–5 greenish-grey, with brown spots. Mid May to June. Incubation 12–13 days; fledging 12–13 days. One brood.

Spotted Flycatcher

*Below:* Pied Flycatchers take readily to boxes. *R. Hosking/ FLPA*

**Pied Flycatcher** *Muscicapa hypoleuca*
Summer resident but rather local and absent from south-east England. Particularly in oak woods and alder and birch woods along rivers or streams. Catches insects in flight by hawking, but also takes them from trees and on the ground. Worms and berries occasionally.

Nests in tree holes, walls. Bark, leaves, grasses with a lining of fibres and grass.
Nestbox: Enclosed, with entrance hole 1⅛in–2in (29mm–50mm) in diameter, inside depth not less than 5in (125mm), floor not less than 4in × 4in (100mm × 100mm). Have a convenient perch close to the nestbox, but not on it. Takes readily to boxes, which seem to supply a real need.
Eggs: 4–9 pale blue. Mid May. Incubation 12–13 days; fledging 13 days. One brood.
Read: *The Pied Flycatcher*, Arne Lundberg, Poyser, 1992.

**Long-tailed Tit** *Aegithalos caudatus*
Resident and generally distributed except in very barren districts and islands. Thickets, bushy heaths, coppices and hedgerows. Also woods in winter.
Feeds in trees, sometimes on ground, restlessly searching for insects and seeds. Parties visit gardens and bird tables for suet, pudding, bread crumbs, grated cheese, etc, especially in hard weather. Peanut enthusiasts once they get the taste.
Nests in bushes, furze or brambles, sometimes in trees. Large egg-shaped nest of moss woven with cobwebs and hair with a lining of many feathers. Entrance hole near top.
Eggs: 8–12 eggs, sometimes unmarked, sometimes a cap of spots or freckles. March/April. Incubation 14–18 days; fledging 15–16 days. Normally one brood.
Read: *Titmice of the British Isles*, J. A. G. Barnes, David & Charles, 1975; *British Tits*, Chris Perrins, Collins, 1979.

**Marsh Tit** *Parus palustris*
Resident and widespread in most of England and Wales, but not Scotland or Ireland. Deciduous woods, hedgerows, thickets and sometimes in gardens. Likes to be near woodland and not, as might reasonably be imagined from its name, in marshes.
Forages over trees for insects; on ground for weed seeds, beechmast, berries and sunflower seeds. Comes to bird table and hanging devices for food as Blue Tit.
Nests in holes in willows, alders, sometimes in walls. Moss with lining of hair or down.
Nestbox: As Blue Tit.
Eggs: 7–8 white, spotted red-brown. End April and May. Incubation 13 days; fledging 16–17 days. Generally one brood.

**Willow Tit** *Parus atricapillus*
Resident. Fairly frequent in parts of south-east England, scattered locally elsewhere. Marshy or damp woods, hedges and thickets.
Forages over trees and on ground for insects, spiders and berries. Will come to bird table for seeds, peanuts.
Excavates a nest chamber in soft rotten wood – usually birch, willow, alder or Elder. Pad of down mixed with wood-fibre, some feathers.
Nestbox: As for Blue Tit, but not enthusiastic. Stuff it full of sawdust or polystyrene chips so that the Willow Tit has to excavate a hole. But a far more successful method is to get a rotten Silver Birch or alder trunk about 6ft (1.8m)

long and 5in or 6in (125mm or 150mm) in diameter, and strap it to a convenient tree, allowing the bird to finish the job. Cap the top with polythene so that rain cannot penetrate easily. It seems that the presence of a suitable rotten tree which they can excavate is all that is needed to attract them to breed in an area they visit during winter. Nests are usually between 2ft and 5ft (600mm and 1.5m) high, averaging 3ft (900mm), so place the trunk accordingly. Birch is the preferred site, alder and Elder are a poor second.
Eggs: 8–9 white, spotted brown-red. Late April and May. Incubation 13 days; fledging 17–19 days. Probably one brood.

**Crested Tit** *Parus cristatus*
Resident in a few parts of north-east Scotland only. Mostly found in pine forests and woods.
Forages mainly on tree trunks for insects, ripe pine cone seeds, berries. Will come to feed at tit-bell and, sometimes, at bird table. Dutch birds patronise red-mesh peanut bags.
Nests in holes or crevices in old and decayed pine stumps, also in alders and birches and sometimes in fencing posts. Dead moss lined with hair of deer or hare, sometimes feathers or wool.
Nestbox: Enclosed, with 1⅛in–1½in (29mm–38mm) entrance hole, interior depth not less than 5in (125mm), floor not less than 4in × 4in (100mm × 100mm).
Eggs: 5–6 white, splotched with chestnut red. End April and May. Incubation 14–15 days; fledging 17–18 days. One brood.

**Coal Tit** *Parus ater*
Resident and generally distributed. Wooded country and gardens with a preference for conifers. Not so commonly found in orchards and hedgerows.
Forages in trees, especially conifers, for insects and spiders. On ground, for seeds and nuts. Not quite so common at bird tables as Great and Blue Tits, but will take the same foods.
Nests in tree, wall or bank holes, close to ground. Moss with thick layer of hair or down and feathers.
Nestbox: As Blue Tit.
Eggs: 7–11 white, with reddish-brown spots. Late April and May. Incubation 17–18 days; fledging 16 days. Sometimes two broods.

**Blue Tit** *Parus caeruleus*
Resident and generally distributed except in north-west Scotland. Woodland, hedges, gardens.
Forages in trees, hedgerows and around houses. Eats wheat, nuts, seeds and insects. Damage to buds and ripe fruit outweighed by consumption of insects. Pugnacious, will hold insect prey with its feed and dismember with bill almost like a hawk. Confiding species that will come readily to bird feeding stations for almost anything. Hauls peanuts 'beak over claw' in a version of the natural behaviour involved in pulling leafy twigs closer to inspect for caterpillars. Milk-drinker, as Great Tit.

There are less than a thousand breeding pairs of Crested Tits in Britain. They are confined to the Scottish Highlands, where the RSPB is clearing alien trees to make more room for the Scots Pines which provide them with homes and a living. Skilful wielding of the chain saw is creating more 'natural' nest-hole opportunities. *Top: B.B. Casals/ FLPA; Below: Tony Soper*

*Opposite:* The tit family; colour variations in the plumage of closely-related species.

Great Tit
Blue Tit
Coal Tit
Long-tailed Tit
Marsh Tit
Willow Tit
RG

Nests as Great Tit. Blue Tits may go to a nestbox because the best natural sites have been taken by the dominant Great Tits.
Nestbox: Enclosed, with 1in–1⅜in (25mm–35mm) entrance hole, otherwise as Great Tit.
Eggs: 7–14 (though there is a record of 19!), usually spotted light chestnut. Late April and May. Incubation 13–14 days; fledging 15–21 days. One brood. Blue Tits breed most successfully in deciduous woodland, where there is an abundance of caterpillars. Their breeding success is least in built-up areas, even though their clutch sizes are smaller to compensate for the poor food available.

**Great Tit** *Parus major*
Resident and generally distributed, scarcer in northern Scotland. Woodland, hedges, gardens.
Forages in trees and hedgerows for insects, spiders, worms. Fruit, peas, nuts and seeds. Does some damage to buds in spring, but it was once estimated that one pair of Great Tits will destroy 7,000–8,000 insects, mainly caterpillars, in about 3 weeks. Fierce bird that will attack and eat a bee. Comes freely to bird table, to hoppers and scrap baskets, where it will display its acrobatic powers as it takes coconut, peanuts, hemp and other seeds, meat, fat, suet, pudding and cheese. May help itself to cream off the top of your milk bottle if you leave it too long on the doorstep.
Nests in tree or wall holes, or crevices. Also in second-hand nests, or the foundations of larger nests. If no natural sites are available, it may use letterboxes, flower pots, beehives and almost any kind of hole. Nest lined with a thick layer of hair or down.
Nestbox: Enclosed, with 1⅛in (29mm) diameter entrance hole or slightly larger, interior depth at least 5in (127mm) from hole to floor, and floor at least 4in × 4in (100mm × 100mm). Great Tits are the most enthusiastic customers for boxes, with Blue Tits coming second. They not uncommonly occupy the same box, the Great Tits taking over, covering the Blue Tits' eggs with a fresh lining and hatching only their own eggs (though mixed broods are not unknown). Tit boxes are successful even in woodland, if there is a shortage of old trees because management procedures do not tolerate them. The result is a strong competition for suitable sites.
Eggs: 5–11 white, splotched reddish brown. End April to June. Incubation 13–14 days; fledging about 3 weeks. One brood.

**Nuthatch** *Sitta europae*
Resident and fairly common in Wales and southern England. Old trees in woods, parkland, gardens.
Dodges about on tree trunks. Wedges nuts, acorns, beechmast and seeds in crevices, and hacks them open with bill. Also takes insects. Will come freely to bird table and hanging devices for hemp, seeds, nuts, cake, fat, etc. Try jamming a brazil nut into a crevice.
Nests in tree holes or sometimes in holes in walls. Female fills crevice and reduces entrance to desired size with mud. Lines nest with bark flakes or leaves.

## A WEDGE-SHAPED NESTBOX FOR TREECREEPERS

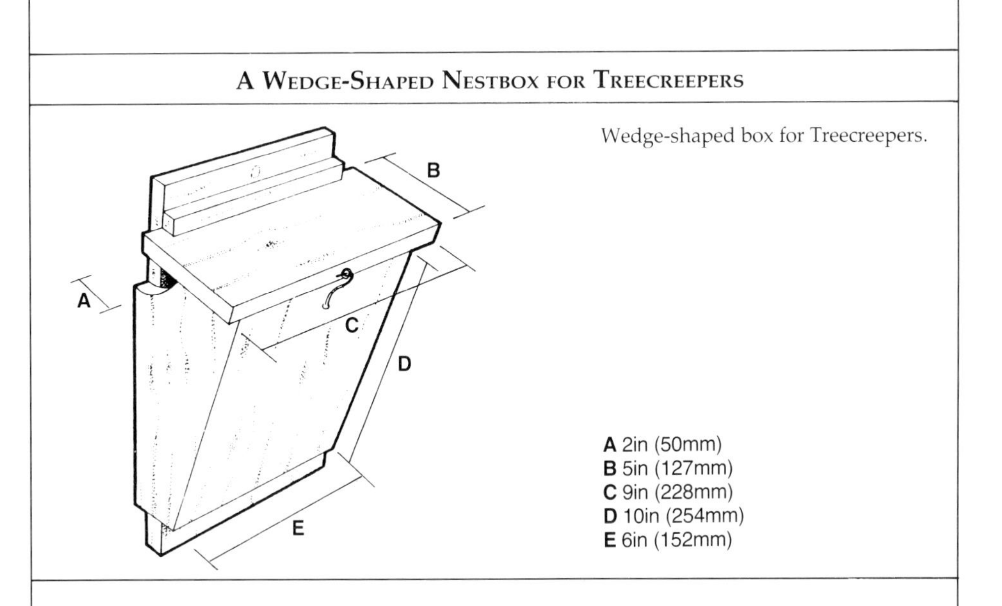


Wedge-shaped box for Treecreepers.

Nestbox: Enclosed, with 1⅛in–1½in (29mm–38mm) entrance hole, interior depth not less than 5in (127mm), floor not less than 4in × 4in (100mm × 100mm).
Eggs: 5–9 white, spotted with red-brown. End April to June. Incubation 14–15 days; fledging about 24. One brood.

**Treecreeper** *Certhia familiaris*
Resident and generally distributed. Woodland, parks, gardens with large trees.
Forages unobtrusively for insects over trees. Does not come to bird table, but may indulge in crushed nuts, porridge or suet fat spread in crevices of rough-barked trees, especially Wellingtonia. Has visited peanut feeders. Try putting out uncooked pastry.
Nests behind loose bark or cracks on tree trunks, or behind Ivy. Sometimes in wall or building crevices. Twigs, moss, grass, lined with feathers and bits of wool.
Nestbox: May come to conventional enclosed type, but a wedge-shaped box has been specially designed with their needs in mind, though I've had one up for years without success. An alternative design involves a book-shaped box 7¼in (180mm) tall by 4¾in × 1⁹⁄₁₆in (120mm × 40mm), with a 2in × 1in (50mm × 25mm) entrance hole at the top of the 'spine'. Clamp it to a tree trunk at around 10ft (3m) high. Or try securing a loose piece of bark to a tree trunk to simulate a natural crevice. Entrance must be alongside tree trunk.
Eggs: Usually white, with red-brown spots at larger end. End April to June. Incubation 14–15 days; fledging 14–15 days. There may be a second brood.

**Jay** *Garrulus glandarius*
Resident and generally distributed. Woodland, never far from trees.
Hops about branches and on ground. Mostly vegetable food, peas, potatoes, corn, beechmast, nuts, fruit and berries. Animal food includes eggs and small birds, mice, slugs, snails, worms and insects. Eats large numbers of acorns and,

Nuthatch

like other crows, has the habit of burying acorns and other surplus food in secret places in trees and under ground. Shy bird, except in some well-timbered suburban areas where it becomes very tame and will come to the bird table or ground station for almost any food. Has learnt to shake spiral peanut-holders to dislodge nuts in order to pick them up from the ground.
Nests fairly low in undergrowth or tree-fork. Sticks and twigs and a little earth, lined with roots and perhaps hair.
Eggs: 5–6 sage-green or olive-buff eggs, mottled with darker olive spots. Early May. Incubation 16–17 days; fledging 20 days. One brood.

**Magpie** *Pica pica*
Resident and generally distributed in England and Wales, scarce in parts of Scotland. Farmland and open country with hedges and trees; suburbs and urban areas.
Frequently in pairs or small parties foraging on ground and in hedgerows for insects, small mammals and birds. Cereals, fruit, nuts, peas and berries. Will come to bird table or ground station for large scraps, which it takes away. Fond of the milk bottle. Not to be too warmly welcomed because of its predatory habits in the breeding season but its impact on songbird populations is less than popularly believed.
Nests in tall trees, thorny bush or neglected hedgerow. Bulky, domed structure of sticks, with an inner lining of earth and roots.
Eggs: 5–8 eggs, greenish-blue to yellowish and greyish-green, spotted and mottled brown and ash. April onwards. Incubation 17–18 days; fledging about 22–27 days. One brood.
Read: *Crows of the World*, Derek Goodwin, Natural History Museum, London 1986; *Crows*, Franklin Coombs, Batsford, 1978.

**Chough** *Pyrrhocorax pyrrhocorax*
Resident. Rare, but a stable and healthy population of less than a thousand pairs in the British Isles. Confined to Scotland, Wales and Ireland (the Celtic Crow, sadly none left in Cornwall now).

Magpie

Jackdaw

Nests in crevices or holes in cliffs and sea caves (though often inland in Wales, in disused quarries and in mineshafts). Bulky structure of sticks, heather stalks, etc. lined with wool and hair. Many breed in man-made features such as ruined buildings, castles, bridges, lighthouses, etc.
Has been known to use artificial covered nest sites in ruins and such places as Martello towers.
Eggs: 3–4 blue with brown spots. Late April onwards. Incubation 17–18 days; fledging about 38 days. One brood.
Read: As Magpie.

**Jackdaw** *Corvus monedula*
Resident and common except in north-west Scotland. Farm and parkland, cliffs, old buildings.

Jaunty bird, feeding in parties or flocks on animal and vegetable matter. Will take young birds and eggs if it gets the chance. Comes freely to bird table or ground station for scraps, cereals, potato, fruit, berries and nuts. Fond of macaroni cheese.
Nests in colonies in trees, buildings, rocks or Rabbit burrows, holes, cracks or crevices. Almost any hole will do – often in bottom of Rook or heron's nest. Twigs, sometimes very bulky, sometimes not. Lining of grass, wool, hair, etc.
Nestbox: Enclosed type, with not less than 6in (150mm) entrance hole, 17in (430mm) interior depth, and at least a 7½in × 7½in (190mm × 190mm) floor. Or open type as for Kestrel.
Eggs: Usually 4–6 pale greenish-blue, spotted brownish black. Mid April. Incubation 17–18 days; fledging 17–18 days. One brood.
Read: As Magpie.

**Rook** *Corvus frugilegus*
Resident and generally distributed. Agricultural areas with trees for nesting.
Feeds openly on ground in small parties or flocks. Cereals, potatoes, roots, fruit, nuts, berries, insects, worms. Will also feed on carrion (dead lambs, etc), and kill small birds in hard weather. Comes freely to ground station for almost anything.
Nests in tree-top colonies normally. Mass of sticks solidified with earth, lined with grasses and straw.
Eggs: 3–5 light blue-green to green and grey-green eggs. Late March onwards. Incubation 16–18 days; fledging 29–30 days. One brood.
Read: As Magpie.

**Starling** *Sturnus vulgaris*
Resident and generally distributed. Found almost anywhere, having successfully adapted to Man's ways.
Active bird, foraging on ground and in trees and hawking for insects. Animal and vegetable foods of almost any kind. Enthusiastic bird table and ground station visitor. Sometimes defeated by hanging devices, but individuals have even learnt to extract peanuts from mesh bags. Very fond of fresh creamy milk and leg-of-lamb bones, particularly marrow, but will eat anything available.
Nests, often in colonies, in tree or building holes. Untidy structure of straw and grasses lined with feathers.
Nestbox: Enclosed, with entrance hole 2in (50mm) diameter, inside depth 12in (300mm), floor area 9in × 9in (230mm × 230mm).
Starlings will explore many possibilities of piracy, and will sometimes take over an old tit box, when the wood has softened enough to enable them to hack away at the hole and enlarge it. If they annoy you by taking over the nestbox, consider that there may be a Great Spotted Woodpecker nearby who has been spared eviction.
Eggs: 5–7 pale blue. End of March onwards. (They often get taken short and lay one on the lawn.) Incubation 12–13 days; fledging 20–22 days. Usually one brood, two in south-east England.
Read: *The Starling*, Chris Feare, Oxford University Press, 1984.

**House Sparrow** *Passer domesticus*
Resident and widely distributed. Cultivated land and vicinity of human habitation.
Operates in non-territorial 'gangs', cleaning up wherever there are easy pickings on farms, hedgerows, parks, gardens, docks, railways, and 'waste' land of all kinds. Corn, seeds, insects. Tough customer at the bird table, eating almost anything, especially cereal-based foods. Wastes a great deal. Much too successful; where there are too many House Sparrows other birds tend to get crowded out. Has learned to extract peanuts from net bags, sometimes even hovering to do so. May hang upside-down tit-style to get at nuts from a 'difficult' feeder.
Nests in holes or niches around occupied houses: eaves, drainpiping, creeper, also in hedges and trees, House Martins' nests, or in the foundations of Rooks' nests. Untidy structure of straw and grasses lined with feathers and oddments. In cramped locations may consist of lining only.
Nestbox: Enclosed, with entrance hole 1¼in (32mm diameter, inside depth not less than 5in (125mm), floor area 6in × 6in (150mm × 150mm).
May easily become a pest, denying nestboxes to more welcome birds. Drastic solution is to destroy nests as soon as they are built.
Eggs: 3–5 greyish-white, finely spotted grey and brown. May to August. Incubation 12–14 days; fledging 15 days. Two to three broods.
Read: *The Sparrows*, J. D. Summers-Smith, Poyser, 1988; *In Search of Sparrow*, J D. Summers-Smith, Poyser, 1992.

**Tree Sparrow** *Passer montanus*
Resident and widely distributed in England, Wales, eastern side of Scotland and a few parts of Ireland. 'Country cousin' of House Sparrow, frequenting same habitat but less attached to human habitations.
Feeds on weed seeds, corn, insects, spiders. Will visit bird table for seeds and scraps but is a shy bird compared with the House Sparrow.
Nests in holes of trees, banks, haystacks and thatch, buildings and in foundations of disused Rook or Magpie nests. Untidy, similar to that of House Sparrow.
Nestbox: Enclosed, with entrance hole 1⅛in (28mm) diameter, inside depth not less than 6in (152mm), floor not less than 4in × 4in (100mm × 100mm). Very susceptible to disturbance.
Eggs: 4–6, smaller, browner, darker than those of House Sparrow. Late April to August. Incubation 12–14 days; fledging 12–14 days. Two broods usually.

**Chaffinch** *Fringilla coelebs*
Resident and widely distributed. Gardens, hedgerows, woods, commons, farmland.
Forages on ground and in trees. Insects, spiders, fruit, fruit buds. Tame and enthusiastic bird-tabler, taking seeds of all kinds, bird pudding, scraps and berries.
Nests in hedgerows, orchards, gardens, not choosy. Beautiful structure of moss with interwoven grass and roots, decorated with lichens held together by spiders' webs. Lined with hair and feathers.

*Above:* Starlings.

*Opposite:* Finches.

Brambling
Male
Female
Male
Greenfinch
Male
Chaffinch
Female
Female
Female
Siskin
Male
Redpoll
Goldfinch
Male
Bullfinch
Female
Male
Female
Reed Bunting
RG

Eggs: 4–5 greenish-blue to brownish-stone eggs, spotted/streaked purplish-brown. Mid-April to June. Incubation 11–13 days; fledging 13–14 days. Mostly single-brooded.

**Brambling** *Fringilla montifringilla*
Has been known to take mixed seeds dropped from bird table and peanuts from a mesh bag. Visits bird table in hard weather, or when beechmast crop fails.

**Greenfinch** *Chloris chloris*
Resident and common. Gardens, shrubberies, farmland.
Feeds sociably on ground and in trees. Seeds of all kinds, berries, fruit tree buds, occasionally beetles, ants, aphids. Comes to bird table and seed hoppers for sunflower seed especially, but is most enthusiastic about peanuts. Will appear where not previously seen when peanut bag hoisted. Will even eat buckwheat. Berries of yew, Ivy, hawthorn, Elder, etc. Windfalls.
Nests in hedgerows and evergreen bushes and trees. Moss interwoven with twigs and lined with roots and hair, sometimes feathers.
Eggs: 4–6 eggs, ground colour dirty white to pale greenish-blue, variably spotted red-brown. Late April/May onwards. Incubation 13–14 days; fledging 13–16 days. Two broods.
Read: *Finches*, Ian Newton, Collins, 1972.

**Goldfinch** *Carduelis carduelis*
Resident and generally distributed. Gardens, orchards and cultivated land.
Small flocks flitter around plant seed-heads, not so much on the ground. Seeds, especially of thistles, Teazle and other weeds. Also insects. Will come infrequently to the bird table for small seeds of grains and grasses. Crack some hemp for them, as their beaks are not so strong as those of other finches. Have been known to take peanuts from a string.
Nests especially in fruit trees and chestnuts. Also in hedges and thick berberis. Elegant nest of roots, grass, moss and lichens, lined with vegetable down and wool, placed far out at the end of the branch.
Eggs: 5–6 bluish-white eggs, spotted and streaked red-brown. Early May onwards. Incubation 12–13 days; fledging 13–14 days. Two broods.

**Siskin** *Carduelis spinus*
Resident in parts of Ireland, Scotland and Wales and in Devon, the New Forest and Norfolk. Increasing. Mainly winter visitor, widely distributed. Woods in summer, otherwise copses, streams, gardens.
Seen in mixed parties with Redpolls searching spruce, birch and larch for seeds. Since the mid-1960s has become increasingly common in gardens in winter, a habit which spread from Surrey through the south-east. Perhaps first attracted by suitable seed-bearing trees, it has stayed to enjoy the bird table, specialising in meat fat and peanuts. Very tame, seeming almost indifferent to Man, though aggressive in behaviour to other birds. Said to be especially attracted to peanuts in *red* mesh bags; though one observer found that while red mesh failed, nuts in a *white* RSPB scrap cage did the trick.

Nests in conifers, high up. Moss and wool interwoven with grass and twigs. Lined with rootlets, down and feathers.
Eggs: 3–5 eggs. April to May. Incubation 11–12 days; fledging about 15 days. Two broods.

**Linnet** *Acanthis cannabina*
Grain-eater, may come to bird table for kitchen scraps.

**Redpoll** *Acanthis flammea*
Grain-eater, may come to bird table for kitchen scraps, especially in Scotland.

**Crossbill** *Loxia curvirostra*
Late summer visitor. Varying numbers. Every few years invades and over-winters in great numbers, many individuals remaining to breed. Coniferous woods, gardens and parks. Clambers about branches parrot-fashion in parties, wrenching off pine and larch cones. Holds cone in foot while it splits the scales and extracts the seed with its tongue. Apart from cone seeds, will eat thistle seeds, berries and insects. Very tame, it will visit bird table for seeds, especially sunflower. Very fond of water and bathing.
Nests on pine branches. Foundation of twigs, cups of moss, grass and wool lined with grass, fur, hair, feathers.
Eggs: 4 greenish-white eggs with few spots/streaks of purple-red. January to July. Incubation 12–13 days; fledging more than 24 days. One brood.

**Bullfinch** *Pyrrhula pyrrhula*
Resident and generally distributed. Shrubberies, copses, gardens, orchards, hedgerows.
In autumn and early winter eats mainly weed seeds, some berries; in a hard winter, if its natural food, ashmast, is short it will ravage fruit tree buds. Remedy is to spread 'Transweb' (for address see page 186) (*Transatlantic Plastics Ltd)* among branches of smaller trees and shrubs. Unsightly but effective. Not keen on bird tables, may occasionally come for seeds and berries, but is especially fond of black and red rape. Will take peanuts from a mesh bag (or from another bird), but cannot extract them from shell.
Nests in hedges, evergreen bushes, creeper, brambles. Foundation of twig and moss, cup lined with interlacing roots and hair.
Eggs: 4–5 green-blue eggs with few purple-brown spots and streaks. Late April onwards. Incubation 12–14 days; fledging 12–17 days. At least two broods.

**Hawfinch** *Coccothraustes coccothraustes*
Resident, generally distributed, but not much in evidence; local in Great Britain, but very rare in Ireland. Woodland, parks, orchards and wooded gardens.
Feeds in trees, taking kernels and seeds. Fond of green peas. Will come shyly to bird table for fruit, seeds and nuts. Highly-developed bill muscles enable it to crack cherry and plum stones, etc, to extract the kernel.
Nests on fruit tree branches or in bushes and other trees. Foundation of twigs supports shallow cup of lichens, moss, grass lined thinly with roots and hair.

Eggs: 4–6 eggs, ground colour light bluish or greyish-green spotted and streaked blackish-brown. Late April onwards. Incubation 9½ days; fledging 10–11 days. Occasionally two broods.
Read: *The Hawfinch*, Guy Mountfort, Collins, 1957.

**Snow Bunting** *Plectrophenax nivalis*
Has patronised Scottish bird tables, and vessels of the Royal Navy in northern latitudes.

**Yellowhammer** (Yellow Bunting) *Emberiza citrinella*
Resident and generally distributed. Farmland with hedgerows or bushy cover, bushy commons and heaths. Very common along roadsides.
Feeds mainly on ground, hopping and pecking for corn, weed seeds, wild fruits (including blackberries, which most birds don't like), leaves, grasses. Insects, spiders, worms, etc. Will come to scattered 'Wildbird Trill' type seed or to a garden seed-hopper once it has discovered it (as will Cirl Bunting, incidentally), but not an enthusiastic garden bird.

Nests in bottom of hedgerow or bush. Straw, grass, stalks, moss-lined with hair and grass.
Eggs: 3–4 eggs, whitish to purplish to brownish-red with dark brown hairlines and spots. Late April to August or later. Incubation 12–14 days; fledging 12–13 days. Two or three broods.

**Reed Bunting** *Emberiza schoeniclus*
Resident and generally distributed except in Shetlands. Reed-beds, rushy pastures, marginal land and hedgerows, having expanded recently to add dry country and suburbia to its ancestral wetland habitat.
In wintertime joins with Yellowhammers and finches in open fields and visits gardens, often in early spring, for seeds and crushed oats.
Nests in marshy ground, with thick vegetation, sometimes in bushes. Dried grasses and moss, lined with fine grasses, hair.
Nestbox: May take advantage of goose/duck raft and nest on it.
Eggs: 4–5 eggs, bluish. May/April. Incubation about 13–14 days; fledging 10–13 days. Two or three broods.

## BREEDING SEASONS OF GARDEN NESTING BIRDS

| Bird | March | April | May | June | July | August | |
|---|---|---|---|---|---|---|---|
| Stock Dove (2–3) | | | | | | | |
| Collared Dove (2–4) | | | | | | | Oct |
| Turtle Dove (2) | | | | | | | |
| Cuckoo (up to 15–16 eggs) | | | | | | | |
| Tawny Owl (1) | | | | | | | |
| Rose-ringed Parakeet (1) Jan | | | | | | | |
| Swift (1) | | | | | | | |
| Great Spotted Woodpecker (1) | | | | | | | |
| Swallow (2–3) | | | | | | | Sept |
| House Martin (2–3) | | | | | | | Oct |
| Pied Wagtail (1–2) | | | | | | | |
| Wren (2) | | | | | | | |
| Dunnock (2–3) | | | | | | | |
| Garden warbler (1) | | | | | | | |
| Blackcap (1–2 in south) | | | | | | | |
| Whitethroat (2) | | | | | | | |
| Lesser Whitethroat (1) | | | | | | | |
| Willow Warbler (1) | | | | | | | |
| Chiffchaff (1) | | | | | | | |
| Goldcrest (2) | | | | | | | |
| Pied Flycatcher (1) | | | | | | | |
| Spotted Flycatcher (1–2) | | | | | | | |
| Robin (2–3) | | | | | | | |
| Blackbird (2–4) | | | | | | | |
| Song Thrush (2–3) | | | | | | | |
| Mistle Thrush (1–2) Feb | | | | | | | |
| Long-tailed Tit (1) | | | | | | | |
| Marsh Tit (1) | | | | | | | |
| Coal Tit (1) | | | | | | | |
| Blue Tit (1) | | | | | | | |
| Great Tit (1) | | | | | | | |
| Nuthatch (1) | | | | | | | |
| Treecreeper (1) | | | | | | | |
| Chaffinch (1) | | | | | | | |
| Greenfinch (2–3) | | | | | | | Sept |
| Goldfinch (2–3) | | | | | | | |
| Linnet (2–3) | | | | | | | |
| Bullfinch (2–3) | | | | | | | |
| Hawfinch (2–3) | | | | | | | |
| House Sparrow (2–3) | | | | | | | |
| Tree Sparrow (2) | | | | | | | |
| Starling (1) | | | | | | | |
| Magpie (1) | | | | | | | |
| Jackdaw (1) | | | | | | | |
| Carrion Crow (1) | | | | | | | |

**Notes:**

**1.** Breeding season as shown is taken from *average* dates of first laying (south of England) to approximate fledging of last brood. For seasons in northern England and Scotland start one week to ten days later.

**2.** The usual number of broods reared in a season is shown in brackets. There will be many exceptions; for example in good years single-brooded species may suddenly nest twice; some individual pairs of single-brooded species (Blue Tit, Great Tit) may produce two broods in any given season.

## ONE OR TWO SURPRISES

**Rose-ringed Parakeet** *Psittacula krameri*
This African/Indian species has been colonising the fringes of London since first escaping from captivity (or being deliberately released) around 1969. Now breeding successfully and spreading. Suburban parks, large gardens. Feeds freely at bird tables and enjoys fruit. Offer it dates, if you can afford them. Nests in tree holes. Presumably it will take to enclosed boxes.

**Budgerigar** *Melopsittacus undulatus*
Escaped or released cage birds sometimes breed in the wild in the south east. There was once an established free-flying colony on the island of Tresco, in the Scillies. These birds, often very tame, will come readily to feeding devices for seeds.

**Canary** *Serinus canaria*
As Budgerigar, will use seed feeders.

6

# BIRDING FROM THE KITCHEN WINDOW

## ESTABLISHING A HIERARCHY

A well-founded bird table will be a delight to watch, especially in winter when it is fulfilling its purpose. A continuous stream of finches and tits will share the pickings with Starlings, sparrows and perhaps even woodpeckers and Nuthatches. There will be much interest in putting names to the birds, in sorting out their plumages, and in watching the spotty juveniles adopt their adult flying suits. But there will also be interest in watching behaviour at the bird table. Some species will appear tolerant of all-comers, some will object strongly – even to the company of their own kind. Both Robins and Blackbirds, for instance, may hold their breeding territories through the winter, and do not encourage trespass. Blue

*Below:* Pecking order at the bird table. This Blue Tit is likely to see the Long-tailed Tit off.

*Above right:* half-coconuts provide good fatty food and only the more acrobatic birds, like this Great Tit, find it easy to get at it. *Walter Murray/NHPA*

Birds actively defend their breeding territory. Any trespassers are fiercely repulsed, even if they are reflections in a car mirror. This behaviour also commonly occurs in the face of reflective hub caps or windows and it wastes a great deal of a bird's energy. *Eric & David Hosking/FLPA*

Tits and Greenfinches, by contrast, enjoy a daily round where they may visit a whole series of different gardens. There is a suggestion that some Great Tits are developing a tendency to sit tight in garden territories, possibly encouraged by the very availability of bird table food.

Study the behaviour of your breakfast guests over a period of time, and you should be able to work out a pecking order. Birds are easily inclined to quarrel over their food, and these feeding squabbles are well observed at the bird table, where they inevitably come into close contact. One threatens another by posturing, ie gaping aggressively, spreading wings and tail. It is largely a game of bluff, since neither individual wants to come to blows, wasting energy and risking the loss of precious feathers, to say nothing of the danger from predators if they aren't keeping a proper watch. But the game has a serious object, because the winner gets the choicest titbit (and at other times, the best perch, the best breeding territory and the most desirable mate). So the establishment of a pecking order is a meaningful affair, and it plays a real part in everyday bird life, to say nothing of our own.

The peck order, or, more scientifically, dominance hierarchy, is so called because the experimental work which demonstrated its validity was carried out with domestic hens. They establish dominance by pecking about the head and shoulders of rivals. It applies to species which live social or colonial lives, involving a great deal of shoulder-rubbing with other birds, not necessarily of their own species. The process involves fights, bickering and bluff which continues until an order emerges. From the boss bird downwards, everyone knows his place, though bickering is constant, with individuals jostling and 'trying it on' with the object of improving their rating. The dominant cock has it all his own way, eating the best food and fathering the most chicks on the most attractive hens. He therefore leads an aggressive life, defending and consolidating his position till he is inevitably toppled as age creeps up on him. In a mixed flock there will still be a peck order, which explains why the 'greedy' Starlings take precedence at the bird table, followed in the hierarchy by House Sparrows, Great Blue, Marsh or Willow and Coal Tits in that order. In fact Blue Tits will rob Great Tits almost as often as they are robbed by them, but without a doubt these two species dominate the other tits, with Coal Tits decidedly the weakest in the hierarchy.

Blackbirds fight enthusiastically among themselves, with a great deal of noisy chasing across the lawn and through the shrubbery. As a species they dominate Song Thrushes, taking earthworms from their beaks and waiting till they crack the meat out of snail shells before moving in to steal it. If you are lucky enough to have a resident Mistle Thrush, he will dominate both Blackbirds and Song Thrushes! Starlings are aggressive and quarrelsome by nature; it is one of the traits which has led them to success, in population terms at least if not in terms of our approbation. This can be seen as they work over a garden lawn, but is most obvious at the bird table, where the constraints of space and the stimulus of abundant food work them up to fever pitch. Working fast, they grab the biggest bits and, in their anxiety to fill up and get away from a potentially dangerous situation, they scatter food far and wide. In early spring, the sexes are easily distinguishable, the males having blue-grey at the base of the bill where females show pink. Armed with this knowledge you will soon see that the males are the ones feeding on plenty at the bird table, while the submissive females are banished to less attractive places.

## PREDATORS

One of the advantages of communal activities, as practised by Starlings, is that there is safety in numbers, and predators are less likely to take advantage of surprise where many are feeding. For birds have good reason to be apprehensive while foraging at the bird table: however carefully you have sited it, and however effectively you have protected them from land-based predators like cats and Weasels, they are at risk from potent and persistent enemies, airborne raptors belonging to their own class – owls and hawks.

When there is a sudden hush in your garden, with the small birds dashing to disperse in cover, calling only the sharp cries of alarm, there is a predator about. And the small birds are well able to recognise their enemies. A Robin will cringe when a Sparrowhawk passes, but take no notice at all if a goose flies by. There is good reason for alarm, and the necessary information is programmed into the Robin at birth. This innate knowledge tells which creatures to ignore and which to run away from, information which is the result of lessons learnt long ago. Doubtless, the immediate experience of seeing one of your kind killed reinforces the understanding in a powerful manner. And, seeing the encounter from the other viewpoint, it is equally true that the Sparrowhawk carries programmed information about suitable prey species. He has a built-in 'search image' which may encourage him to specialise in Blue Tits because instinct tells him that Blue Tits are for catching. And Sparrowhawks are very efficient at catching small birds. It has been shown that they may take two-and-a-half per cent of a whole Chaffinch population in the month of May.

The raptors are well designed for their job as hunter-killers. Sparrowhawks have broad, rounded wings by comparison with the more open-country falcons like Peregrines, which have been designed for speed. But Sparrowhawks work in amongst the trees and hedgerows, and while they enjoy a fair turn of speed they are also able to engage in fast turns and complicated manoeuvres, a useful facility if you're chasing a wildly jinking small bird. Like the other birds of prey, they are able to turn their fourth toe so that it is pointing backwards, allowing a tight grip with two sharp-clawed toes on either side of the victim's body. And the forward component of the impact motion as they land on a victim causes the claws to lock automatically and grip fast, a sinister variation on the same mechanism which locks songbirds' feet to their roosting perch when they go to sleep. The bird must make a conscious effort to release its victim.

Sparrowhawks are not the only hunters which enjoy the living bounty of the bird table. Kestrels commonly take House Sparrows and young Starlings when they get the chance, behaviour that has been most observed in London, where

Sparrowhawks are natural predators in a bird garden, so try not to begrudge them their prey.

they are now well established as breeding birds. Voles are their preferred diet, and they find good hunting along railway embankments, but if small mammals are scarce they will take many birds, and a bird table will become more interesting to them. Perhaps more surprisingly, there are records of Kestrels coming to take broken dog biscuits and uncooked bacon rind from an Edinburgh bird table in a cold winter. The fact is that birds like Sparrowhawks and Kestrels are increasingly becoming aware of the potential offered by the bird feeding stations. No fewer than eight species have been recorded at their dirty work. Apart from those already mentioned there have been reports of Tawny, Barn and Little Owls; Merlins; Buzzards; and, astonishingly, Goshawks. The Goshawk was seen taking small birds from a bird table in a Yorkshire garden which adjoined a forestry plantation – typical habitat.

However disconcerting and, perhaps, upsetting it may be to see your garden birds carried off struggling by a predator, it is a perfectly natural everyday event in the bird world. Certainly it is not our responsibility to try to put an end to this hunting by controlling predators, which was the automatic, albeit ignorant, reaction of gamekeepers in the days when they held the view that anything which ate their precious charges must be doing harm. (It would also be naive to think that all present-day gamekeepers have seen the light, but there is some evidence that a large proportion of them have!) To begin with, the predator is not as all-powerful as he is sometimes seen; he doesn't kill every time and, indeed, once he is unmasked he is molested by the very birds he is seeking. Blackbirds chivvy cats, Rooks chivvy Buzzards and songbirds gather to chivvy owls the moment they reveal themselves. This mobbing is a form of display, the birds most at risk banding together to draw attention to the danger. Hoping to avert attack, they feel there is some safety in numbers.

It then follows that the prey taken by a predator is going to represent the slowest/dimmest/weakest/most disabled of its stock, and therefore the stock is improved by this weeding out of less healthy individuals which might have bred and passed on their weakness. Finally, it is generally accepted that it is not in the predator's long-term interest to reduce its prey species. For example, the Sparrowhawk has a special interest in ensuring there will be an abundance of sparrows for its progeny to chase. There is a nice parallel here with the behaviour of the foxhunting fraternity who, however often they may say they are trying to exterminate Foxes, are in fact concerned to conserve their numbers in a comfortable balance so that there will always be enough to chase.

Thus, there is no sense in taking the law into your own hands and shooting or trapping predators – it doesn't serve the interests of the prey. Other potential natural enemies will step in to take advantage of the available surplus, or disease or parasites or shortage of food will perform the predator's task instead. And there are always far fewer predators than the animals on which they feed. Songbirds need only a relatively small patch of land to supply their needs, but a hawk or an owl needs a hunting territory which will run to dozens of acres. Insects are always more numerous than the shrews which eat them, and shrews are always more common than the owls which in turn eat *them* – a demonstrable concept which is known as the 'pyramid of numbers', where the so-called 'higher' animals come out on top. In essence an effective check is kept on the predator's population, in that it can only survive in relation to a higher population of its preferred prey species.

Sadly the vision of a paradise garden where all is happiness and serenity is an unacceptable concept to the biologist. Predators, whether Sparrowhawk or beetle, will do their thing and make their presence known. But it is all a question of attitude, and surely we should glory in the knowledge that their function is a totally healthy one. They leave the garden a stronger place than they found it. And their method of control is far preferable to the sort that comes from a bottle or a box bought at the garden centre.

## PREDATORS AND POISONS

A bird's life is fraught with natural hazards. After surviving a cold winter, it may get snapped up by a Sparrowhawk. If it succeeds in finding a mate and hatching young, there may be a sudden shortage of food and the weaker nestlings may die. The chances of a wild bird living to a ripe old age are so remote as to be almost non-existent. So if we are going to invite birds to join us in our gardens there is an obligation on us to try to reduce the hazards, while recognising the fact that the predator-prey relationship is basically a healthy one, with advantage for both parties.

You might want to seal your garden completely from unwelcome predators, but this is more easily said than done. Sparrowhawks and Weasels are all part of the natural scene (Weasels are adept at raiding nestboxes), but domestic cats and Grey Squirrels are less acceptable. Ideally, there is no place for them in the bird garden, but you will never keep them out, short of total war. If you have a cat of your own, at least it serves the useful purpose of discouraging alien cats in defending its territory. And you might like to consider keeping it in for a reasonable and regular period each morning and before dusk, in order to give the birds time to feed. I have to admit that I like the company of a cat and would not care to be without one, but do not try to *tame* your garden birds if you keep a cat. Feed them and make homes for them by all means, but don't encourage them to become too friendly or there will, inevitably, be a tragic outcome.

If you don't want to go as far as installing a wire fence, the best substitute is a thick and prickly hedge. Hawthorn or Holly hedges will both in time become fairly impenetrable, although you will always have to watch for secret passageways and block them with bramble or thorn cuttings. The disadvantage of a clipped hedge is that it will not fruit very freely, though, on the other hand, it provides good nesting sites. Allow some of the plants to mature so that a few trees grow out of the hedge to blossom and fruit. Holly is particularly good, because the dead leaves cover the soil underneath with spiny points which may deter cats, Weasels and such like.

Rats have to be taken seriously. They climb well, even shinning up trees and hedges to search for eggs and young birds, and a good bird garden is also an attractive rat garden. So food should not be left on the ground at night, and windfalls too should be cleared away every evening: they can become part of the winter bird table menu.

Grey Squirrels, too, are unwelcome visitors to the bird garden. They may appear charmingly acrobatic as they leap from branch to branch, but they are great egg-eaters. They will even enlarge the hole of a nestbox to lift out the nestlings. You must harden your heart and shoot or trap them. You might want to discourage Jays and Magpies, for there is no doubt that they will take any small birds' eggs and young they can find. On the other hand, they are handsome birds in their own right, and there is as yet no evidence that they affect garden populations adversely.

For information on pest control without poison, contact the Henry Doubleday Research Association (address page 183).

## Sick, Injured and Orphaned Birds

Birds may fall into your hands for any number of reasons, accidents or cat-mauling being the most likely. The law requires you to return them to the wild as soon as possible. First, be sure that they are in need of help. In the breeding season young birds just out of the nest are inclined to wander about the garden apparently 'lost'. This highly vulnerable stage in their lives is a perfectly natural one, their parents are never far away, and your best plan is to leave them in peace and hope for the best. If birds do need help (and apart from the juveniles described above, if you can pick them up they are certainly in trouble) here are some simple hints:

- Sick birds need overnight quiet, warmth and darkness first. Identify the species, if possible.
- Appropriate food should be given after an initial period of rest.
- Diseases or injuries should be treated by a veterinary surgeon.
- Birds beyond complete recovery should be destroyed humanely.
- Do not attempt to clean oiled birds; contact the RSPCA.
- Young birds should usually be left undisturbed.
- Birds must be released as soon as possible.

A sick bird or an orphan should be handled as little as possible and should not be made into a pet. Some birds can become tame very easily and it will be difficult for them to be released; remember that only a completely healthy bird can survive in the wild. Caring for any sick creature involves considerable time, money and patience and if you are not prepared to take on this responsibility, or if the bird is injured beyond recovery, it is better to be realistic about this and have the bird humanely destroyed rather than allow it to suffer from neglect.

Please note that the RSPB is not able to deal with problems involving birds in captivity – its charter confines it to consideration for *wild* birds and their environment. But, in collaboration with the RSPCA, it has produced a useful leaflet *Care of Sick, Injured and Orphaned Birds*, available free from any RSPB regional office or from its headquarters (for address see page 182).

**Further reading:**

*Care for the Wild*, W. J. Jordan and J. Hughes, 1988.

## UNUSUAL GARDEN VISITORS

One of the pleasures of bird table watching is the slow but steady way in which more and more species are being lured to join in. Not only Kestrels and Goshawks, fortunately, but more and more of the birds which formerly kept us at some distance. Great Spotted Woodpeckers and Long-tailed Tits are much commoner visitors than they once were. Goldcrests, Firecrests, Cirl Buntings, Woodcock, Common Snipe, Water Rail, Kingfishers and Dipper – all are species which have taken to the habit. Bearded Tits have been encouraged to visit seed piles near reed beds at the Minsmere RSPB reserve. Little Gulls have taken food scraps from a litter bin. It seems there is no limit to the possibilities.

Where you live will, of course, have a considerable bearing on your potential bird list. Sailors at sea have operated bird feeding stations for many years. On migration, many land birds have sought respite on weather ships and oil rigs, as well as ships steaming at sea. Kestrels which visited a weather ship at station India stayed three days and were fed on steak. A Little Auk was once desperate enough to take tinned sardines, Bramblings and Snow Buntings, at 71° and 74° North respectively, enjoyed 'Swoop'. A Great Blue Heron landed on the flight deck of HMS *Hermes*, of Falkland fame, when she was 500 miles east of Puerto Rico in the Caribbean, and was entertained on the quarterdeck for two days with dishes of Pilchards. The cargo vessel *Sugar Crystal* was joined 140 miles south-west of Ireland, while on passage to Felixstowe, by thirty-four Jackdaws and three Rooks, which squabbled over grain spillage on the decks. On a more everyday level, no cross-Channel yachtsman can have failed to have the company, at some time or another, of a tired racing pigeon looking for a rest and some sustenance.

Great rarities have first revealed themselves on the bird table in the severe winter of 1954–5, while there was thick snow on the ground, an unfortunate American warbler, which had found itself in England after an unscheduled Atlantic crossing, showed up on a south Devon bird table. This Yellow-Rumped Warbler, the first record for Britain, established itself in vigorous ownership of the food supply, seeing off the resident Blue Tits and even going so far as to deprive one of a feather, a considerable loss to a small bird in such freezing conditions. The warbler displayed a keen interest in bread and marmalade, thus causing some of us to suggest that it came from Peru to look for Paddington Bear.

The first record of a Red-throated Thrush, *Turdus ruficollis*, to be seen in Britain was made in a north Buckinghamshire urban garden in the winter of 1978–9. A Siberian species, this one was looked at with a certain amount of suspicion since it seemed highly possible it was an escaped cage bird. In the same way it would be unwise to assume that the Budgerigars spotted on some bird tables had flown in from Wagga Wagga, Australia. But the Naumann's Thrush which visited gardens in London in February 1990 was a genuine Asian vagrant.

Rarities and reluctant arrivals aside, there are long-term trends to be discerned by studying the bird table. Great Spotted Woodpeckers and Long-tailed Tits have come to stay, and the colonisation of our islands by Collared Doves has been made easier by the freely available supply of food. Sparrowhawks and Kestrels will presumably take to bird table visiting in increasing numbers.

Not surprisingly, crows have learnt the advantages of artificial food. Magpies, as part of a general increase in numbers, have become more common in country gardens as well as the suburbs, where there is a scarcity of gamekeepers. First they took the songbirds' eggs and young, and then moved in to take the food provided for them. Though it may be something of a strain for the bird enthusiast to see Magpies culling his hard-won neighbours, one just has to sit back and let them act as healthy predators and sort things out amongst themselves. So far the evidence is that they do not affect songbird populations adversely.

Jays will collect unshelled peanuts and carry them off for burying and subsequent digging up as a food store, in the same way that they bury vast numbers of acorns in autumn. This food storage is typical of other species such as Nuthatches and the tits, which hide food at a time when it is abundant, though it is particularly widespread amongs the crow family. It occurs when there is more food about than the birds are able to eat, and quite apart from obviously suitable items like nuts, they will hide bread or cheese pieces. This family propensity to store food gave rise to the 'thieving Magpie' legend. And while wild birds do not carry off gold rings and diamond necklaces, it is highly likely that tame ones might do so when they are deprived of their natural foraging.

In flight the Song Thrush reveals a yellowish underwing. The Mistle Thrush's tail is edged white at the tip while the Fieldfare's tail is black.

Mixed flocks of thrushes often visit gardens to feed on berries or fallen apples. The differences in size are then seen more clearly, although they fluff themselves up in cold weather.

## WINTER FEEDERS

Birds derive advantage from the autumn plenty by putting on fat and, as we have discussed in some cases, by laying up a winter store. It is in autumn, when berries and seeds are plentiful, that you are at least likely to have a well attended bird table – the natural food available is a greater attraction. But it may be that your bird table is not particularly successful in a winter which follows a bumper seeding season. In other words, there is no substitute for natural food, and birds will prefer it given a choice. As well as seeds, there will still be a measure of life-support in the hibernating flies, spiders, woodlice, centipedes and so on, which find just enough warmth to overwinter in the fallen leaves of undergrowth. One of the characteristic sounds of winter is the crunching noise made by Blackbirds as they thunder about in the shrubbery, foraging.

It is in hard weather, when the extreme cold requires more energy output by birds to maintain their body temperatures, that the bird table is a life-saver. The most obvious effect of the arrival of a cold snap is that more species will come for the food. There will be Mistle Thrushes, Greenfinches, Long-tailed Tits, and more Blackbirds. Fieldfares and Redwings will visit the garden lawn and the bird table, and there will be Bramblings feeding under the nut bags for fallen morsels. In extreme conditions, even the fiercely territorial Robins will feed side by side. Other species will find their way in from exposed country to enjoy the relative shelter, including Reed Buntings and Yellowhammers, Grey Wagtails, Skylarks and Meadow Pipits, Pheasants and Moorhens. Wrens may take crumbs from bird tables in a way that is entirely untypical.

It is perhaps true that the existence of bird table food has made it possible for some normally migrant birds to stay with us and stick out our British winter. Blackcaps, which are overwintering in increasing numbers, mostly in the mild south and south-west, are particularly vulnerable to severe winters. They rely heavily on berries, such as those of Cotoneaster and Honeysuckle, will eat the Holly berries which are not exactly popular generally, and turn to Ivy berries during the early spring as their preferred food. Rotting windfall apples are also important to them, and they have even been seen to take Mistletoe berries, a fruit which seems of little interest to most birds apart from thrushes. It is in hard weather that Blackcaps are most likely to be seen at bird tables, looking for cake crumbs, fat, nuts and seeds, and this is the time when a tray of rejected fruit from the shop will be most welcome. Overwintering Chiffchaffs will come to the bird table, too, for crumbs and suet.

It is possible that bird table offerings have fuelled the range expansion of an exotic invader, the Rose-ringed Parakeet. This attractive looking parrot, originating from Africa and India, escaped from captivity, or was perhaps deliberately released in some numbers (as a human response to its unrewarding behaviour in captivity), at the end of the 1960s. Since then, starting from a nucleus in the London suburbs and Kent, it has slowly but surely colonised the south-east and established itself as a feral species and something of a pest. Omnivorous by nature, it prefers fruit and has a devastating effect on apple orchards with its tendency to take just a couple of pecks at each fruit. It comes freely to bird tables and, breeding in tree holes, will presumably take to nestboxes. Even the most severe winters have failed to halt its spread, in spite of knowledgeable predictions that it could not survive the cold. Broadly speaking, it is true that birds are perfectly able to withstand the lowest temperatures, provided they are well fed and protected by healthy plumage.

In extreme winter conditions, when water is frozen and mudflats glazed by ice, water birds and waders suffer greatly. Herons and Kingfishers must go down to the sea to find open water; Pied Wagtails cannot find their ditchside insects; thrushes and Robins cannot penetrate the frozen earth. And small birds like Long-tailed Tits, Wrens and Goldcrests may die because there aren't enough feeding hours in the day for them to meet their high energy requirements. So, once you start operating a winter bird table, it is of the utmost importance that you keep it well supplied through the dark months until April.

## TOOLS OF THE TRADE

Birdwatching is a modified form of hunting. Primitive hunters sought only to fill their bellies, but if we are to enjoy success in our terms we must use their techniques in our aim of getting close to the quarry, both in the literal sense and in that of getting to know them better. To this day the hunter can teach any aspiring birdman a great deal. Fieldcraft involves a great deal more than wearing a battledress jacket. The hunter knows his prey as well as he is able, he moves quietly and with due regard to the wind and the light, above all he knows time and place. Present-day wildlife photographers face all the problems of the hunter, and it is axiomatic that the best ones are those to whom a knowledge of natural history comes first, before knowledge of film stocks and photo apparatus.

The tools of the trade are important. First and foremost are those which are standard issue to all, or at least most of us. Keen eyesight and hearing are the naturalist's most precious assets, followed by an ability to use them effectively. Observation by sight and sound is the basis of all fieldwork. Perhaps the observer's notebook and pencil come next for the best memory in the world is no substitute for on-the-spot recording. Most introductions to birdwatching contain useful sections on field recording, but perhaps the most practical, enjoyable and stimulating is that contained in Ian Wallace's excellent book *Discover Birds* published by Whizzard Press/André Deutsch in 1979. Wallace takes the reader on a wild romp to most of Britain's best bird places, and his enthusiasm is both constructive and infectious.

Binoculars are well-nigh indispensable, and they are a source of great heart-searching to many would-be birders. The problem is that, like birds, they come in a bewildering variety of guises. To meet all eventualities, you need to own half-a-dozen pairs. However, given that you are to start by buying one pair, you should go for glasses that are reasonably light in weight and which give a bright picture over a medium field of view, with a magnification of 8, 9 or 10 times and objective lenses between 30mm and 50mm in diameter. Thus your chosen binoculars might be described as 8 × 30 or 10 × 40, both highly suitable everyday combinations. Lower or higher magnifications should go hand in hand with higher light-gathering power, eg 7 × 50 or 10 × 50, the first being ideal for marine use, when you must reduce the problems of unwanted rock and roll, and the second about as powerful as most people can hold without getting a wobbly picture. But see what the people in your local bird club use – you'll learn a lot simply by borrowing their glasses on your first outing or two.

Zeiss 10 × 40 BGAT binoculars are just about ideal for all-round birding. *Tim Soper*

Avoid glasses with magnifications greater than 10 as the plague, until you are experienced enough to use them to advantage. Avoid zoom binoculars, which will be heavy, and avoid anything which presents you with a blurred image anywhere in the picture or which exhibits colour halo effects. Get the opinions of other birdwatchers and get hold of a field guide pamphlet called *Binoculars and Telescopes* published by the BTO (for address see page 182).

Once you have chosen the binocular power which suits you, the best advice is to buy the most expensive you can afford. But during the trial stage it's probably best to enjoy the relatively cheap 'Avocet' glasses marketed by the RSPB. Write to the Sales Dept (for address see page 182) for a copy of the Sales Catalogue and for membership details. Membership of the Society not only gives your soul the warm glow of satisfaction induced by the knowledge that you are making a highly practical contribution to the well-being of British birds, but brings you a number of benefits, not least of which is the excellent quarterly magazine *Birds*. Local centres, run by the RSPB, provide opportunities to join field excursions, to hear top-class speakers and to see the best of the bird films. It goes without saying that you should also join your local bird club, in order to get the benefit of meeting people who have a strong interest in the bird patch which you share. Local knowledge is vital and is freely shared by genuine birders.

Whether your interest is in far-flung ornithological expeditions or in studying the action nowhere more demanding than your back garden through the kitchen window with a mug of coffee in your hand, you need to be able to make a positive identification of the birds you see or hear. There are several field guides to identification – hunter's manuals – and you must choose the one which suits you best. There are three which are the most widely used. Peterson, Mountford and Hollom's *Field Guide to the Birds of Britain and Europe* was first published, by Collins, in 1954 and it has enjoyed high approval by birdwatchers. The current edition is completely revised and up to date. It has stood the test of time, and can only really be faulted on its tiresome pagination and layout, which are not designed to make for quick reference. Another Collins field guide, *The Birds of Britain and Europe, with North Africa and the Middle East*, by Hermann Heinzel, Fitter and Parslow, published in 1972, is perhaps not to be recommended as a first purchase since it covers a wider geographical area than a beginner birdwatcher ought decently to be interested in. *The Hamlyn Guide to the Birds of Britain and Europe* by Bertel Bruun, published in 1978, has the great advantage of a conventional layout which marries text, maps and drawings so that they appear on the same page for each bird. Examine all these three before you choose.

As for general introductions to bird biology and behaviour, there are excellent books by David Saunders (*RSPB Guide to British Birds*, published by Hamlyn, 1975), by Peter Conder (*RSPB Guide to Birdwatching*, Hamlyn, 1978), by James Fisher and Jim Flegg (*Watching Birds*, T. & A. D. Poyser, 1974), and there is even my own offering (*Birdwatch*, David & Charles, 1991).

You will also need a more encyclopaedic volume of background information species by species, and this is well catered for by *The Popular Handbook of British Birds*, P. A. D. Hollom, Witherby, 1988.

The magazine *Bird Watching* appears on the newstands every month and is full of practical advice and up-to-date news. And you will want to keep informed with the birdwatchers' monthly magazine *British Birds*, a lively and authoritative journal (for address see page 183). Write to Mrs Erika Sharrock, Fountains, Park Lane, Blunham, Bedford MK44 3NJ for a free sample copy. And when your birdwatching develops into an unquenchable craving for more knowledge and you initiate your own research you must share your results with the rest of us by submitting them to *British Birds* for publication. But I warn you, it is about as easy to get a 'note' into *British Birds* as it is to get a letter accepted for *The Times* correspondence page.

Books are notoriously useless at imparting information about the sounds made by birds. Yet their calls and songs are often a vital clue to their identification, to say nothing of their state of mind. Probably the best way to learn the birdsong is by way of a knowledgeable companion. But a good gramophone record or tape will help. Jeffery Boswall has edited a splendid set of vocalisations recorded by Sture Palmer in *A Field Guide to the Birdsongs of Britain and Europe*, which is available (from the RSPB, for instance) in both disc and cassette form. John Kirby's *Wild Track* cassettes are also available from the RSPB. *British Bird Songs and Calls* is a set of 109 recordings compiled by Ron Kettle and available from the Natural History Book Service (see page 183) in cassette or compact disc form.

## BIRDS AND THE LAW

All wild birds, their nest and eggs are protected by law, and, with certain exceptions, it is an offence to kill, injure or take any wild bird. For more information write to the RSPB for a copy of the free leaflet *Birds and the Law* (for address see page 182).

## BIRD TOPOGRAPHY

Putting a name to a bird is arguably the first requirement in getting to know something about it. And a knowledge of bird topography is an important foundation in the building up of identification expertise. Yet unfortunately the charts published in bird identification manuals offer such varied systems that they may confuse more than they help. The editors of the monthly magazine *British Birds* have produced these comprehensive and authoritative 'British Standard' charts in the hope that everyone will use the same language.

CHART OF BIRD TOPOGRAPHY

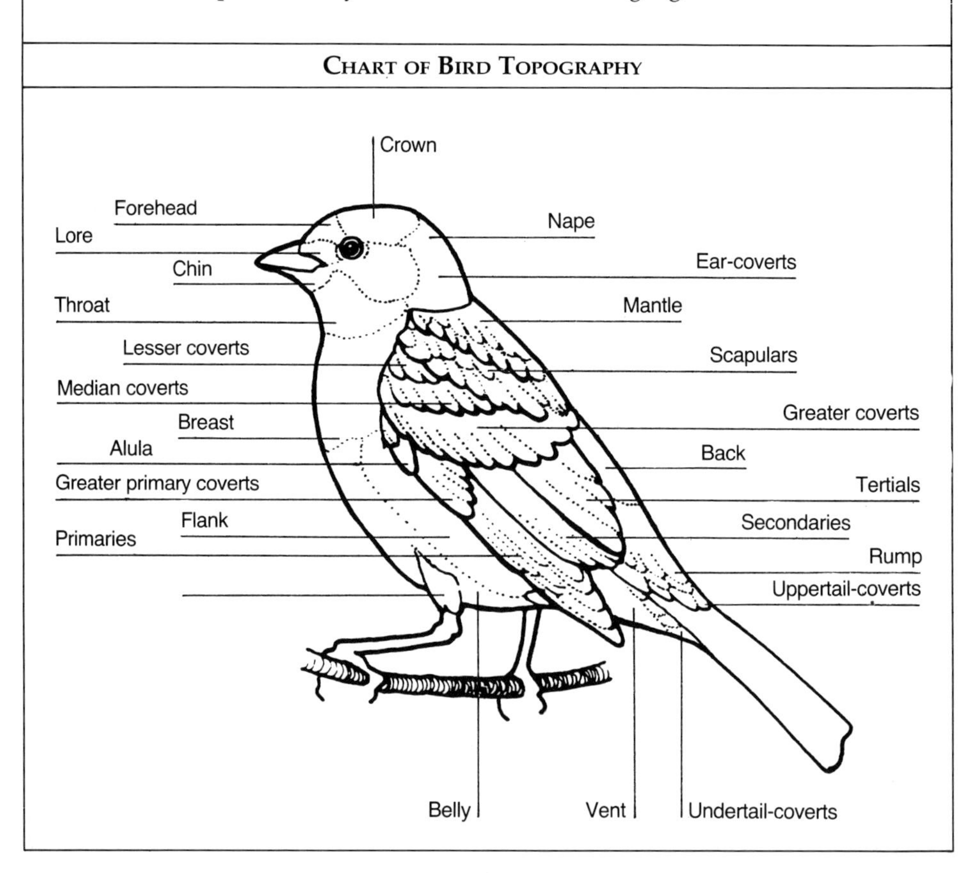

## CHART WITH HEAD AND FOOT DETAILS

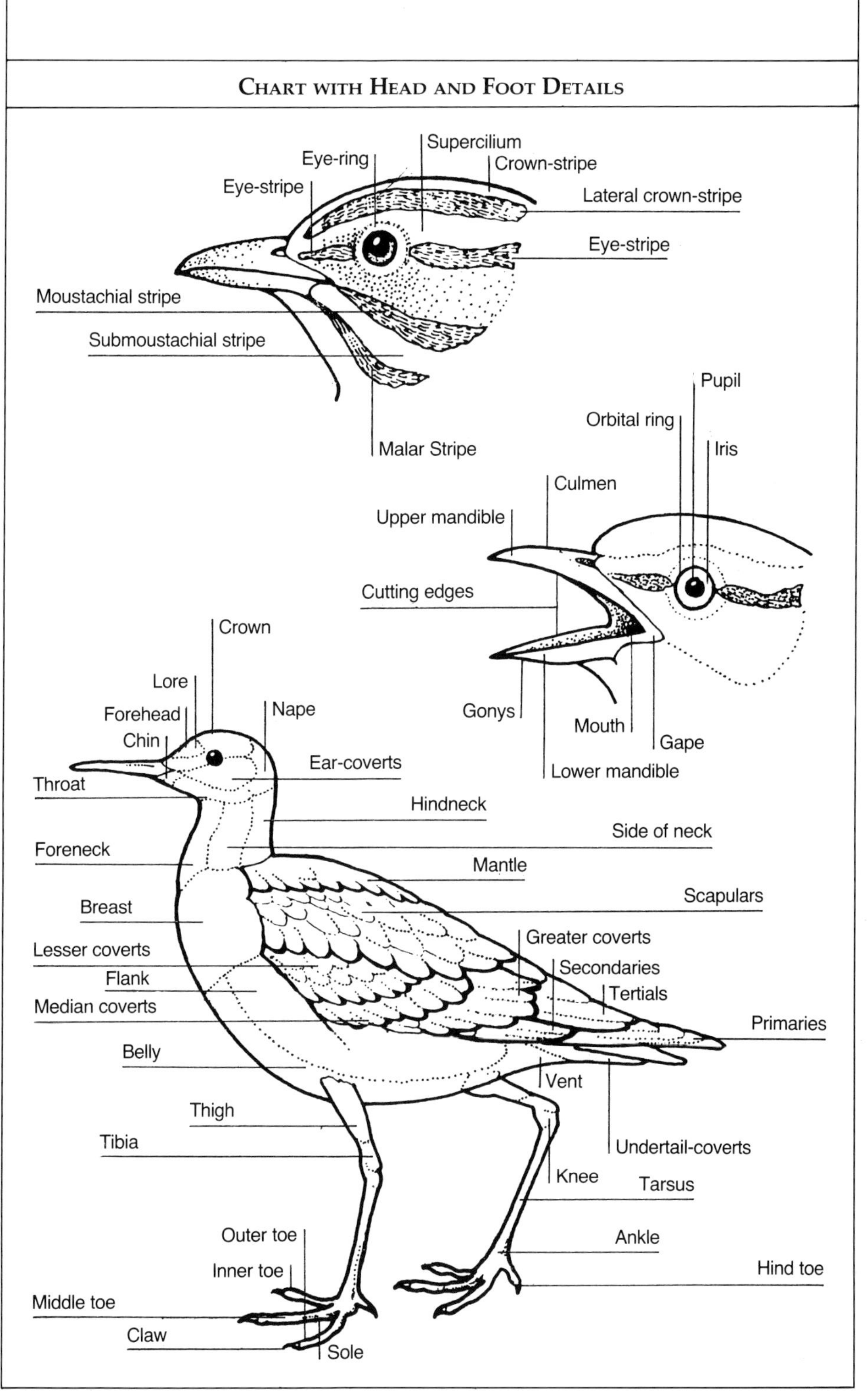

## CHART OF WING TOPOGRAPHY

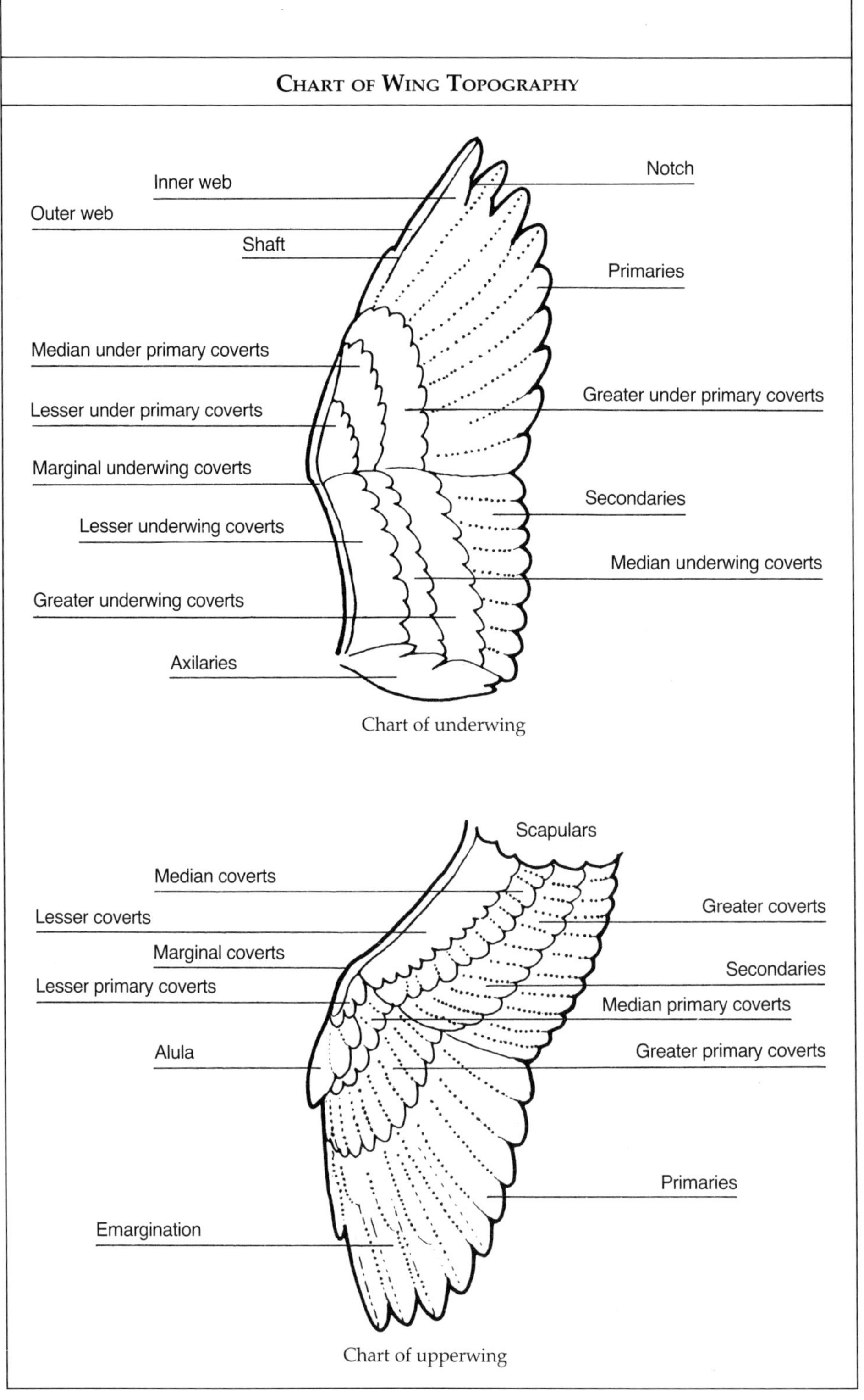


Chart of underwing

Chart of upperwing

7

# WORKING FOR BIRDS

Birds and birdwatching embrace an extraordinary range of potential interests. The scientific study of bird populations and life-styles provides endless material for professional ornithologists, working in a field which covers the gamut from pure academic research to pest control in the everyday world of agriculture. The amateur birdwatcher's interests may range from disciplined census work through a fairly casual enjoyment of a weekend hobby to the highly-skilled pursuit of rarities by the dedicated and single-minded band of 'twitchers', whose goal is to tick off a longer life-list than the next man in a sporting chase. And there are those

The RSPB bird table is well-designed to please the birds. (For supplier see page 184). *Tony Soper*

of us who simply enjoy the Blue Tits on the half-coconut, or who write poetry, prose or music about birds. Their fascination is endless. Much of their charm lies in the way their life-styles run on similar paths to our own, though they enjoy the inestimable advantage of flight capability, the facility which we so greatly envy. Unlike the mammals which are our much closer relatives, birds mostly live in our daylight world of sights and sounds rather than the nocturnal world of smells. They share a great deal with us, including our food, and this makes it easier to identify with them.

Birds are a part of our everyday life whether we like it or not – the sparrows which nest in the roof, pigeons which strut the city streets, gulls which defile the newly washed car, or even some unfortunate bundle of feathers brought in and left by the cat. They are the most easily seen and heard of all our wild neighbours and we are involved in their lives through everything we do, interacting with them and all the other creatures of the community. But many of us want more than this passing interest in birds, we want to record our observations and contribute to the scientific study of birds, even graduating to the role of ornithologist.

The Royal Society for the Protection of Birds, first formed in 1889 and now with the strength of well over half a million members and managing eighty-seven reserves, must take a place of honour among the many organisations which work on behalf of wildlife. The RSPB has research scientists, film makers, wardens, and other specialists – full-time staff earning their living with birds – but their great achievement has been to put their subject across to the general public with enthusiasm through publications, films and meetings.

## LEISURE TIME INVOLVEMENT

What are the opportunities for working voluntarily with birds? After joining the RSPB, probably the best way to get involved in your local area is to join the local Naturalists' Trust and Bird Society. There you'll find out what is going on in the area and meet people who will know all about the national bird organisations. Through them your leisure time hobby can be harnessed for valuable and exciting research.

The British Trust for Ornithology is the major organisation which initiates and co-ordinates the sort of research which involves the efforts of a network of birdwatchers covering the whole of Britain. If you want to become actively concerned with census or ringing work, then you need to become a member. The standards are high, the BTO operates to serious scientific criteria, but the range of work is wide enough to encompass everyone from the conscientious beginner to the dedicated professional. The work tends to consist of some form of census – statistical recording of numbers breeding, passing by and wintering – or ringing. And there is no doubt that this work is invaluable, for birds and bird numbers are among the most sensitive indicators of the health of our environment. Monitoring fluctuations in their numbers provides essential information to those whose job it is to formulate policy in wildlife conservation. And at the same time keeping a record of your bird observations will give you immense interest and satisfaction.

## BTO GARDEN BIRD SURVEY

In the winter of 1970/71 the British Trust for Ornithology launched the Garden Bird Feeding Survey with the main aims of determining which species make use of gardens at different times of year; what range of foods is provided in gardens and which are the preferred ones. Since then bird gardeners and other householders – sometimes with little more than a backyard and a window box – have contributed to the survey from all parts of the country. In the first ten years of the survey alone, 199 different species were recorded in gardens for which observations were sent in! The mind boggles at the gardens which produced Wigeon, Red Kite, Bewick's Swan and an American Blue Jay. What is proved beyond dispute by the survey is the enormous range of species which do, from one time to another, make use of our gardens and the important part which all these gardens play for birds in our ever-shrinking countryside.

Another important fact disclosed is the extent to which the number and variety of birds differs from year to year depending on the severity of the winter. The conclusion is clear: garden bird feeding in winter is a major contributor to the survival of many of our individual birds.

The BTO has now revamped and expanded the survey, generously sponsored by the electronics firm BASF. Relaunched in autumn 1987, the survey now seeks to attract up to five thousand participating garden enthusiasts all of whom are supplied with special recording forms and are asked to record details of the use which different birds make of the garden – not just the feeding station, as in earlier years – right through the year. So if *you* want to become one of Britain's official garden bird recorders write to the BTO (for address see page 182). They will be pleased to hear from you, whether you have a two-acre spread in Sunningdale, a backyard in Halifax or a lighthouse compound in the Irish Sea. But competition to take part in the survey is keen, and it may be that you will have to put your name on a waiting list!

The 'new' Garden Bird Survey not only produces a lot more detail about birds' feeding preferences in gardens but also provides important information on the extent to which different breeding species use our gardens for nesting in different parts of the country. Special emphasis too is placed on one or two species which are currently declining on farmland and which make some use of gardens, for example Linnet and Tree Sparrow, and Magpies which have increased rapidly in suburban gardens in recent years and are accused of undue depletion of garden song birds.

One question which the bird gardener wants answered most is 'What are the commonest garden species?' The results of the years of surveying so far undertaken can give this answer with authority. Although the relative position of each of the commonest species may vary a little from one year to the next, the Top

Twelve are usually as follows in the chart below (based on the percentage of gardens in which they are recorded).

| THE TOP TWELVE GARDEN BIRDS | |
|---|---|
| Blue Tit | Great Tit |
| Blackbird | Greenfinch |
| Robin | Starling |
| House Sparrow | Song Thrush |
| Dunnock | Coal Tit |
| Chaffinch | Collared Dove |

In most years these twelve are followed, a few percentage points behind, by Wren, Magpie, Mistle Thrush, Black-headed Gull and Pied Wagtail. Not surprisingly, with all this bounty of feathered food on hand, the Sparrowhawk comes around twentieth place. Whether or not you take part in the BTO's Garden Bird Survey, keep a list of all the species you see in your garden; it will surprise you and encourage you to go on trying to add to the list.

## NEST RECORD SCHEME

The BTO has designed Nest Record Cards which make it possible to record the success, or failure, of known birds' nests. Submitting completed cards to central office adds to the existing body of statistical information on bird-breeding biology – for instance, the timing of each species' breeding season, the number of eggs laid and young reared, and how breeding success is affected by such factors as climate and human activity. Little detailed information of this kind existed before the BTO began the work of completing cards on a large-scale basis, but by 1992 some 850,000 nest record cards had been completed. The information has resulted in many useful publications involving analyses of these records.

## THE COMMON BIRDS CENSUS

This is, effectively, a scheme which measures the population changes in a given area over a period of years. This enables the maintenance of an annual index of fluctuations in population levels, and makes it possible to discern trends towards change of status. Originally the census was started, in 1961, at the request of the Nature Conservancy Council, with the emphasis on agricultural habitats, since it is on farms that the use of toxic chemicals and hedgerow destruction has had marked effects on bird populations. More recently other habitats have been covered as well, in order to produce more representative results. The work is exacting, but important, and more help is required if we are to know more about the factors which influence the numbers of our everyday birds.

## OTHER SCHEMES

Other work includes the long-running census on heronries, which aims to count the heron nests at a sample of heronries in the British Isles; wader and wildfowl counts through the winter months; the scheme to map the distribution of every species wintering in the British Isles – the Atlas of Winter Birds. All these are organised by the BTO. But most county bird societies (obtain addresses from your local library) have census work of some kind in hand, so it is worth checking with them.

## BIRD RINGING

The seasonal movements of birds have always fascinated us, and we have long sought to unravel their mystery. In part, there is an element of sport in the pursuit, an enjoyment of the challenges offered in trapping and marking the birds, but there is a more serious purpose, that of discovering more about birds' life styles and population dynamics, information which can be of great value in assessing the ecological effects of changes in land use, for instance.

It was in medieval times, somewhere around the twelfth century, that the Prior of a Cistercian monastery in Germany reported that a man who had fixed a parchment to a Swallow's leg asking, 'O swallow, where do you live in winter', received a reply in the following spring. 'In Asia, in home of Petrus'. The Romans had long used Swallows to carry messages to their homes, in the style of pigeon post, but it was in 1740 that Johann Leonard Frisch, a Berliner, attached coloured wool to swallows' legs to discover whether individuals returned to the same nest site year after year. They do!

In the early nineteenth century, J. F. Dovaston, one of the pioneers of field ornithology, repeated this experiment, fastening cello wire round the Swallows' necks. He also attached a copper tag inscribed, in Latin, 'where hast thou gone to from Shropshire?', though sadly he had no returns. Later still, Lord William Percy marked young Woodcock at Alnwick, in Northumberland, with more success, achieving 58 recoveries from 375 ringed birds. But his rings were unnumbered and lacked a return address. The credit for the first use of a bird ring which carried its own unique number, and a return address, goes to a Danish ornithologist, H. C. Mortensen, who ringed 164 Starlings in 1899. In Britain, systematic bird ringing began in 1909, sponsored by H. F. Witherby in London, the founder of the magazine *British Birds*, and A. Landsborough Thompson in Aberdeen. Since 1937 it has been organised by the BTO and only accredited ringers may take part. By 1992, 2½ million birds had been individually marked with rings.

Trapping, for the purposes of ringing, is based on the methods used by hunters through the centuries, though one of the most ancient techniques, that of liming, is thankfully illegal nowadays. At one time Holly bark was stripped in quantities, in the springtime, pounded and mashed by druggists who then supplied the resulting 'birdlime' to hunters. They spread it liberally on suitable roosting-places, where unfortunate songbirds became stuck to their perch. Cage-trapping, decoying, cannon or rocket netting, recordings of bird song – these are all methods used to bring birds to the hand. The most efficient, at the present time, is the use of mist nets made from fine nylon or terylene thread and dyed black. Erected between tall poles to form an almost invisible barrier, they trap any bird which flies into them. The trapped bird may be marked with a brightly coloured dye, which can be seen from a fair distance. However, this method provides only a limited amount of information and lasts only until the wearer moults to a new flying suit. A bird can carry a colour ring which is useful in a limited sense in connection with a small number of individual birds in a relatively restricted area.

Far and away the most useful method of marking is the use of numbered metal rings which carry a return address. In Britain the BTO rings are engraved with the legend 'Brit. Museum London SW7'. Other countries sport their own legends, every ring carrying a unique group of letters and numbers. The records are computerised and stored according to an internationally agreed filing standardisation, and the coverage is almost worldwide. Both ringing and the photographing of birds is governed by the law as laid down in the Protection of Birds Act, and the primary consideration has always been the well-being of the birds themselves.

Once the bird has been ringed and released, the research department of the BTO must sit and wait for subsequent information. Mostly this comes in the form of a letter reporting the bird's death. If, by chance, you come across an avian corpse complete with ring, make a note of the number (send the ring itself, if possible), species of bird (if known) or a description, the place of discovery and the date. Send all this, with any other information you think might be useful, direct to the BTO (for address see page 182). In due course they will respond with details of the bird's ringing history. Pigeon rings, however, should be sent to the Royal Pigeon Racing Association (for address see page 183). If your ringed bird is alive and healthy, simply note the required information and release it!

From ringing records, the BTO has deduced that the roads are a major cause of bird accidents. Twenty-three per cent of Barn Owls meet their end as a result of collision with vehicles. Otherwise predators such as birds of prey, dogs and Foxes account for many, while domestic cats are a major cause of bird death and a helpful source of ring recoveries!

Many and various are the nuggets of information gathered as a result of long-term bird ringing. We know that some birds live long lives, though the average expectancy is very short indeed, especially in the case of songbirds. But an Oystercatcher may live thirty-four years, a Herring Gull, thirty-two. The oldest recorded Swallow covered nearly a quarter-of-a-million miles on its migration journeys during its sixteen years of life; an Arctic Tern, half-a-million in its twenty-seven years. Chris Mead, of the BTO, reckons that the oldest Swift, at sixteen, must have flown over four million miles in its lifetime. In cold weather during the winter of 1963, searching for new feeding grounds, a Redwing covered 2,400 miles in three days. A Swift born and bred in Oxford, was recovered in

Madrid three days after it left the nest. We know that small birds increase their weight just before migration. A Sedge Warbler which weighs about ⅓oz (10g) normally, will build up to more than ⅔oz (20g) thus carrying enough fat to fuel a non-stop flight of 2,000 miles. It completed this distance in just three days, crossing Europe, the Mediterranean and North Africa, possibly even overflying the Sahara to reach Senegal or Ghana, having slimmed down to half its take-off weight and returned to normal.

Chris Perrins trapped Blue Tits at a well-stocked bird table and found that it was visited by more than a hundred different individuals in the course of the morning, while many of us had assumed that our bird table was feeding just the locals.

The mass of information gives muscle to those seeking to influence legislation in a manner which pays due respect to bird requirements both at home and abroad. Birds themselves recognise no political barriers and need to be conserved on a world-wide basis. After all, our Ospreys and Avocets are shot in Spain and North Africa; and our Linnets and Redpolls are trapped in Belgium and France while on passage to these shores. And, at long last, we can map the precise routes of the European Swallows when they leave us to winter in the south. Our British Swallows, for instance, make their way to South Africa, and I have watched them funnelling in to roost in the reed beds of a Johannesburg city park, where the local ringers operate their mist nets.

## OTHER WAYS TO HELP

Finally, if you enjoy watching the birds, but don't want to become involved in field work, there's plenty of scope for voluntary work in fund raising, selling raffle tickets and cards at Christmas – and all the other ingenious ploys devised to provide the necessary finance for bird reserves and their management. Check with your local Bird Society or Naturalists Trust.

# ORGANISATIONS CONCERNED WITH BIRD LIFE

**Royal Society for the Protection of Birds,** The Lodge, Sandy, Bedfordshire SG19 2DL.
Illustrated journal *Birds,* free to members. Manages a network of bird reserves, organises many exhibitions and meetings, is much concerned with conservation and with the enforcement of the Protection of Birds Act. All birdwatchers should support this admirable and efficient society.

**Young Ornithologists' Club**, The Lodge, Sandy, Bedfordshire SG19 2DL. Associated with the RSPB, the YOC is the national club for young people (aged seven to fifteen) who are interested in birds or want to learn about them. Quarterly magazine, *Bird Life,* nationwide projects, outings, courses. Kestrel badge.

**The British Trust for Ornithology**, The Nunnery, Nunnery Place, Thetford, Norfolk IP24 2PU. Major link between amateur and professional ornithologists.
Minimum membership age, fifteen years. Members may take part in organised field studies, ringing and census work. Issues quarterly journal, *Bird Study,* six-weekly *BTO News,* invaluable field guides and other publications. Works closely with RSPB. Lending library. All serious birdwatchers should join. (Send for brochure.)

**The British Ornithologists Union**, c/o The Bird Room, British Museum (Natural History), Cromwell Road, London SW7 5BD.
Senior bird society in Great Britain; its main object is the advancement of ornithological science on a world scale. Quarterly journal, *The Ibis.*

**The Wildfowl and Wetlands Trust**, Slimbridge, Gloucestershire GL2 7BT.
Illustrated annual report and periodical bulletins; maintains unique collection of swans, ducks and geese from all parts of the world.

## OTHER IMPORTANT ORGANISATIONS CONCERNED WITH ALL ASPECTS OF NATURAL HISTORY

**English Nature**, Northminster House, Northminster, Peterborough PE1 1UA.

**Scottish Natural Heritage**, 12, Hope Terrace, Edinburgh, EH9 2AS.

**Countryside Council for Wales**, Plas Penrhos, Fford Penrhos, Bangor, Gwynedd LL57 2LQ.

**Royal Racing Pigeon Association**, The Reddings, Cheltenham, Gloucestershire GL51 6RN.

**Hawk & Owl Trust**, London Zoo, Regent's Park, London NW1 4RY.

**County Naturalists Trusts and Bird Societies**. Addresses are usually available at your local library.

**The International Council for Bird Preservation** (British Section), British Museum (Natural History), Cromwell Road, London SW7.
Issues annual report, co-ordinates and promotes international bird conservation.

**The Henry Doubleday Research Association**, Ryton-on-Dunsmore, Coventry CV8 3LG.
Issues newsletters and occasional publications of interest to the bird gardener. Initiates research into organic farming and gardening methods; pest control without poisons.

**International Union for Conservation of Nature and Natural Resources**, 1110 Morges, Switzerland.
Independent international body promoting and supporting action which will ensure the perpetuation of wild nature and renewable natural resources all over the world.

**The Worldwide Fund for Nature**, 7–8 Plumtree Court, London EC4.
Raises funds and allocates them to projects covering a wide range from land purchase for national parks and reserves to ecological surveys and emergency programmes for the safeguarding of endangered plants and animals.

**The Fauna and Flora Preservation Society**, c/o The Zoological Society of London, Regent's Park, London NW1 4RY.

**The Royal Society for the Prevention of Cruelty to Animals** Causeway, Horsham, Sussex RH12 1HG.

**The Game Conservancy**, Fordingbridge, Hampshire SP6 1EF.

## Other Sources of Information

**British Birds**, the long-established but lively magazine for the ornithologist and serious amateur. Write for a free sample copy to Mrs Erika Sharrock, Fountains, Park Lane, Blunham, Bedford MK44 3NJ.

**Bird Watching**, an up-to-the-minute monthly magazine available from news stands or by subscription from Bird Watching, PO Box 500, Leicester LE99 0AA.

**Natural History Book Service Ltd,** 2 Wills Road, Totnes, Devon TQ9 5XN, holds the most comprehensive stock of world-wide bird books and issues an invaluable seasonal catalogue.

# SUPPLIERS OF FOOD AND EQUIPMENT

## SUPPLIERS OF BIRD FOOD

'Wildbird Trill' is available from most pet stores and supermarkets.

**C. J. Wildbird Foods**, The Rea, Upton Magna, Shrewsbury SY4 4UB. Best peanuts, seeds of all sorts, birdcake mix and 'peckerpacks'. Send for useful catalogue.

**John E. Haith Ltd**, Park Street, Cleethorpes, Humberside DN35 7NF. Wildbird food, soft bill food, songster food, in bulk.

**Jacobi, Jayne & Co**, Hawthorn Cottage, Maypole, Hoath, Canterbury, Kent, CT3, 4LW. Special mix wildbird food, send for list. 'Droll Yankee Birdfeeders', top quality seed and peanut dispensers, exceptionally well made and effective.

## SUPPLIERS OF POND PLANTS, ETC

**Griffin & George Ltd**, Gerrard Biological Centre, The Field Station, Beam Brook, Newdigate, Dorking, Surrey RH5 5EF. Suppliers of pond plants and animals. Send for lists.

**Queensborough Fisheries**, 111 Goldhawk Road, Shepherds Bush, London W12 8EJ. Send for list.

**The London Aquatic Co Ltd**, 42 Finsbury Road, Wood Green, London N22 4PD. Illustrated catalogue.

## SUPPLIERS OF BIRD FURNITURE

**Royal Society for the Protection of Birds**, The Lodge, Sandy, Bedfordshire SG19 2DL. Free catalogue on application. All the RSPB equipment is made of high-quality weather-resistant materials. The woodwork is treated with a harmless preservative and painting is unnecessary. Instructions and advice are included where necessary.

*RSPB Bird Table*, illustrated on page 175, is 18in × 15in (450mm × 375mm), well proportioned. It may be suspended by chains or fixed on to a 2in × 2in (50mm × 50mm) post. Chains and post socket supplied as standard fitments, but *not* the post. The roof overlaps the table slightly and ensures that food is kept dry in most conditions, although strong winds and driving rain inevitably play havoc. But this is the best buy in bird tables. The roof is fitted with rails to take a food-hopper, which slides into place well sheltered.

*Hanging Bracket*, for suspending the table, from a tree, wall or window frame. Strong and rust-resistant.

*Seed Hopper*, for dispensing seed mixtures. Fits neatly under bird table roof, for which it is specially made. Hoppers can be very tiresome gadgets, as they are so susceptible to wind and to rain-jamming.
*Dual-purpose Nestbox*, 8in × 5in × 5in (200mm × 125mm × 125mm). Removable front panel makes the titbox suitable for open-plan occupation by Robins, etc. Metal surround protects entrance hole from squirrels and woodpeckers.

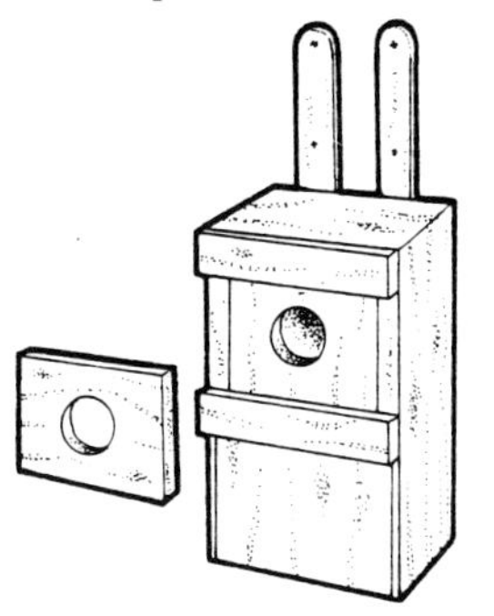

*Nestbox Kit*. Make your own dual-purpose nestbox with this easily assembled kit. 12in × 5in × 5½in (300mm × 125mm × 140mm). Also hanging feeders, nut cages, bird baths, etc.

**C. J. Wildbird Foods**, The Rea, Upton Magna, Shrewsbury SY4 4UB. Free catalogue on application. Well-chosen and tested selection of bird feeders, nestboxes and bird tables.

**Jacobi, Jayne & Co**, Hawthorn Cottage, Maypole, Hoath, Canterbury, Kent, CT3 4LW. *Droll Yankee Birdfeeders*. Top quality seed and peanut dispensers, exceptionally well made and effective. Send for list.

**Nerine Nurseries**, Welland, Malvern, Worcestershire WR13 6LN. Bird tables, tit boxes, Robin boxes and the useful House Martin nests. These act as a magnet to this species and encourage the establishment of a colony. The nests are fixed under the eaves of house or barn. Sometimes House Sparrows may worm their way in by enlarging the holes but Nerine Nurseries supply a leaflet with instructions for a simple anti-sparrow method. Incidentally this method, involving a screen of 12in (300mm) weighted cords, 2½in (63mm) apart, and hung 6in (150mm) away from the entrance hole, can be used to protect natural House Martin nests as well.

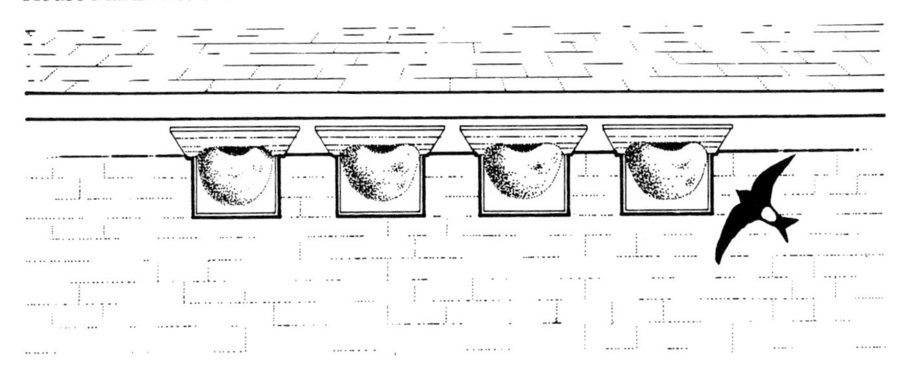

**The Halo Company**, Osborne House, Station Road, Burgess Hill, West Sussex RH15 9EH. Flat-packed corrugated plastic tit-boxes for home assembly. Ideal for field-workers using nestboxes for population studies.

The Sussex Bird Table. *Jamie Wood Ltd* (see below).

**Jamie Wood Ltd**, Cross Street, Polegate, Sussex BN26 6NB. Suppliers of well-made bird tables, bird furniture, hides, etc. SAE for brochure.

**National Trust Shop**, Killerton, Broadclyst, Devon EX5 3LE, for the glazed pottery bird-bell in smart box.

**EcoSchemes Ltd**, Kingmaker House, Station Road, New Barnet, Herts, EN5 1PE. Cavity wall mounting nest boxes.

**Transatlantic Plastics Ltd**, Ventnor, Isle of Wight, suppliers of 'Transweb' which may be used to protect smaller trees' and shrubs' buds from birds.

*Warning*. Avoid any peanut feeder which involves a flexible coiled-spring system. These are highly dangerous, since a bird may get its feed jammed when another flies away from the feeder causing the spring to contract. And avoid all-in-one nestbox/feeding-tray/water/trough devices. It is asking for trouble to invite birds to eat or drink at the doorstep to another bird's house, creating territorial stresses and strains which don't do anyone any good!

# ACKNOWLEDGEMENTS


**Spring 1965**

My grateful thanks go to Mary Blair, Tom Edridge, Elaine and H. G. Hurrell, R. M. Lockley, Winwood Reade and E. H. Ware for ploughing through the first draft of this book and offering so much useful advice. David St John Thomas encouraged me to start; hundreds of BBC Spotlight viewers in the Westcountry forced me to continue (by asking questions, instead of answering them as I had requested). Chris Mead and David Wilson of the British Trust for Ornithology and Frank Hamilton and John Taunton of the Royal Society for the Protection of Birds have been most helpful. Many friends have encouraged me with good ideas and corrections; some have even valiantly improved my grammar, but the mistakes that remain are all my own work.

T.S.

**Spring 1973**

Since this book was first published, I have had a considerable correspondence with readers who have corrected and improved the information in it, and asked yet more questions. My sincere thanks to them; I think I have incorporated all the new material in this edition, but it is clear that no bird gardening book will ever be complete. So I throw in my hand again with a mixture of pleasure and trepidation.

T.S.

**Spring 1977**

At last . . . in colour! My thanks to all correspondents, even those who telephone in the middle of *Dr Who*. Also thanks to Robert Gillmor for his superb drawings and his enviable facility for doing them at the speed of flight, at least when pressure is firmly applied. And to Chris Mead and David Glue at the BTO, who combine scientific standards with an understanding of the enthusiasm of mere pleasure-bird-watchers. And to people like Mr and Mrs Pat Wilson who run the BTO's Garden Bird Feeding Survey, which collects and collates the hard facts. And to Pam Thomas of David & Charles, who makes publishing books fun. I tremble at the thought of what must surely come next – *The Bird Table Book in 3D*. Watch this space!

T.S.

**Spring 1986**

In its twenty-first year *The Bird Table Book* comes of age and reverts to its original title without the whizzkid appellations of 'new' or 'coloured' which decorated it in editions past. But never were truer words written than in my note of 1973 that 'no bird gardening book will ever be complete'.

My continuing gratitude to all those who have taken the trouble to write and to the stalwarts of the BTO who still enjoy birds more than they love their magic computer.

T.S.

**Autumn 1991**
I little thought, twenty-six years ago, when I slaved over a hot pad of foolscap to incubate and hatch and nurture this lusty fledgling, that it would still be alive and kicking through the age of A4 to the word processor. But birds still come enthusiastically to the table and more and more of us welcome them to the feast. Chris Mead's beard is now long enough for the proverbial Wren's nest and Robert Gillmor is President of all the best academies of bird art. The BTO has reached middle age and taken itself to a nunnery, where Chris du Feu presides at the holy nestbox.

Special thanks to Chris Whittle, the sunflower king, and thanks to everyone who has extended my knowledge of the bird garden, and here's hoping that, in time, our cat will change her ways and learn to love baby Robins for their company rather than their crunchiness.

T.S.

# INDEX

Main entries are in **bold type**, illustrations in *italic*.

## A



## B



## C



## D

## L



## M



## N



## O



## P



## Q



## R